Out of the Barn

by

Dick Morley

compiled by Ken Ball

**ISA–The Instrumentation,
Systems, and
Automation Society**

Notice

The information presented in this publication is for the general education of the reader. Because neither the author nor the publisher has any control over the use of the information by the reader, both the author and the publisher disclaim any and all liability of any kind arising out of such use. The reader is expected to exercise sound professional judgment in using any of the information presented in a particular application. Additionally, neither the author nor the publisher have investigated or considered the effect of any patents on the ability of the reader to use any of the information in a particular application. The reader is responsible for reviewing any possible patents that may affect any particular use of the information presented. Any references to commercial products in the work are cited as examples only. Neither the author nor the publisher endorses any referenced commercial product. Any trademarks or trade names referenced belong to the respective owner of the mark or name. Neither the author nor the publisher make any representation regarding the availability of any referenced commercial product at any time. The manufacturer's instructions on use of any commercial product must be followed at all times, even if in conflict with the information in this publication.

This book is published in cooperation with ISA Press, the publishing division of ISA– The Instrumentation, Systems, and Automation Society. ISA is an international, non-profit, technical organization that fosters advancement in the theory, design, manufacture and use of sensors, instruments, computers and systems for measurement and control in a wide variety of applications.

Standard orders may be placed with:

ISA–The Instrumentation, Systems, and Automation Society
67 Alexander Drive   Phone: (919) 549-8411
P.O. Box 12277   Fax: (919) 549-8288
Research Triangle Park,   E-mail: info@isa.org
NC 27709 USA   ISANetwork: www.isa.org

For information on corporate or group discounts, e-mail: bulksales@isa.org.

Library of Congress Cataloging-in-Publication Data in Process

Out of the Barn
Creation and Compilation

O*ut of the Barn* is a compilation of a Dick Morley series of articles which
originally appeared in MSI (formerly *Manufacturing Systems* magazine), then a Thomas Publishing Company monthly periodical. Kevin
Parker was the editor and the major promoter for the series which ran from
1992 into 1997. For his suggestions, encouragement, editing proficiency
and patience with our submitted raw copy, we are most grateful. MSI is
now a Cahners publication and Kevin Parker still serves as editor.

Our thanks and appreciation are also extended to Ken Ball, a friend and
colleague from the early glory days of the programmable controller and our
"Club-of-Detroit." Ken has arranged and assembled the articles into book
form and has added much of the supplemental material. He previously has
been editorial director for *Instruments & Control Systems* magazine, ISA's
InTech and *Programmable Controls* journals and *Industry.Net*.

Acknowledgements & Dedication

The R. Morley Inc. staff is an integrated, extended family group. We work in the Barn (see cover photo), which is a high-tech R&D facility surrounded by several hundred acres of New Hampshire woods. My appreciation goes to the entire staff for an effective and pleasant everyday working environment. The Barn culture and *esprit de corps* has contributed much to my personal productivity and I am forever grateful.

With respect to this book, special recognitions are due my executive administrator, Debbie Morrison, and my creative director (and accomplished daughter), Pat Letourneau. Deb has contributed her professional publishing know-how (including the design and layout of this book), and you'll see evidence of Pat's creative artwork.

Appreciation is also extended to several persons who have made significant contributions to the array of original articles. My late son, Robert Morley; Flavors Technology Marketing Manager Bob DeSimone; and Alan Campagna contributed guest articles. My good friend, Dr. George Markowsky (Computer Science Department Head, University of Maine), gives freely of his time and counsel on just about any topic in our mutual realms of interest.

The book is dedicated to my wife, Shirley, who has been a load-bearing life partner along the tortuous path of our married life. Along with three children of our own, Shirley has handled the heavy end of our being stepparents to some 27 foster kids. She has served as a selectman for our town of Mason, New Hampshire, and has worked winters on the local ski patrol. She rides with me on her own Honda and operates her backhoe for snowplowing and landscaping. My number one accomplishment by far has been marrying Shirley.

—*Dick Morley*

Table of Contents

Foreword

Over the period of January 1992 through August 1997, Dick Morley authored a series of 69 one-page articles for Thomas Publishing's trade magazine, *Manufacturing Systems*. The series was entitled "Under Control" and the articles covered a broad range of topics within the general realm of controls, computers and new concepts in manufacturing. In addition—and typical of Dick Morley—the series included a number of facetious and "off-the-wall" Morley writings which help keep our keels in the water as we sail the uncharted sea of careers.

Dick Morley is widely recognized for his capability to establish bridges between the awesome silicon world and the machinery-dominated factory floor. Along with describing many such ties, the articles also reflect his extensive international business experiences and first-hand knowledge of global manufacturing trends.

After some best-fit iterations, the articles have been grouped under eight topical parts as listed in the contents. As might be expected, many articles did not fit neatly within a single part topic. It is therefore recommended that the book be read as a package wherein various topics are touched upon throughout.

The Morley articles are easy reading and entertaining. Most parts include insights into the technology trends along with comments on business aspects—often from a quite unorthodox viewpoint. However, a greater value may be in the mental challenges Morley imparts as he shares his inimitable analyses and perspectives.

—Ken Ball

"The world of the future

will be an ever more

demanding struggle

against the limitations

of our intelligence"

—*Prof. Norbert Wiener*

Part I:
Floor Level Control Systems

R ecent industrial controls include the widespread use of sensors, Ethernet connections to floor control loops, wireless and embedded computers in machinery and appliances. Many sensors are "smart" with integrated signal conditioning, diagnostics and communications. Smart sensors and terminating devices become intelligent control nodes that lead to distributed control systems.

Programmable Logic Controllers (PLCs) remain the workhorse for the factory floor and are expanding into commercial use. According to Automation Research Corporation's Market Reports, the global market for PLCs in 2000 was over $6.2 billion and is projected to be $7.3 billion by 2005.

"You may try to destroy wealth and find that all you have done is create poverty"

—*Winston Churchill*

Dull, Boring Data Collection?

Data collection has to be the most boring subject ever. How do I psyche myself up to discuss data collection? Bland City! Or so I thought. But as I started to write, I began to see that maybe the subject has some merit after all. Let's give it a shot.

While the main theme of this issue is "Data Collection," the real technology problem today is implementing and maintaining complex systems. Sensing technology is adequate for today's computer and control systems, but changes are coming.

This magazine is *Manufacturing Systems*, not *Manufacturing Components*. We need to look at data collection as the sensory segment of a large, complex enterprise system. The data collection segment of any system has no intellectual function. It exists to supply an image of the world state for the processing and manipulation needs of the integrated enterprise.

Sensors transform raw data into a form compatible with the processing expectations of the computing units. Sensor technologies such as bar codes convert labels into digital information. Using the location of the sensors, digital processors can infer the state of the system.

Often overlooked is the LAN, used to send the sensed and converted information to the central or distributed computers. This is similar to the retina and optic nerve-system of the eye. Raw data (photons) change into a form suitable for brain processing. The retina also compresses the image for transmission over the optic nerve.

These views allow us to predict future trends. Let's take a stab at a few.

We will be using bandwidths of 100 megabits (FDDI—the Fiber Distributed Data Interface standard) in the future to match near-term computer power.

Generic primitive transforms on the data will make standards of sensing applicable to the industry.

Transformation of data will not be to identify "package number" or "temperature – F" but will match LAN requirements.

Linearization, identification and other application-specific sensing will give way to transmission of raw information as described by

We need to look at data collection as the sensory segment of a large, complex enterprise system.

C. Shannon in his 1948 paper on information theory.

When the volume of the data is high, cost is one of the major issues. Jim Pinto of Action Instruments suggests that the Echelon network and his ASIC (Application Specific Integrated Circuit) working together will improve the price/performance of analog sensing by a factor of ten.

Spread spectrum radio, powerline transmission, fiber optics and copper will soon play a strong role.

Sensors themselves will be more reliable and faster than today's offerings. Most of the PLC vendors use the T1 rates (1.544Mb) to collect data. This rate is at least one order of magnitude *too slow*.

As next-generation controls enter the market, next-generation sensors will have to be ready. The next-generation controller (NGC) effort of the Air Force, under the direction of Martin Marietta, will demand new standards of performance. Independent Canadian, German and Japanese advanced controller efforts may eclipse the U.S. effort. Data collection technology will keep pace with these fast-moving events.

Standardization, high-speed generic sensing and the ability to send large masses of data to demanding computers must be the offering of the modern vendor of sensors and sensing systems.

Recently, I was at the supercollider site in Texas. This site requires more than 50 miles of control. The 250,000 points of control require fast conversion and transmission to the processing and control computers. By comparison, a single department in discrete manufacturing may now have 3,000 points of control, mostly binary. Newer process ideas and implementations will require 10,000 points of analog for the same area. Enterprise requirements for high-speed conversion and transmission of more than 100,000 points will be common in the next decade.

Data collection is not the issue, but matching the sensing to the needs of the application is. Success is awareness of the world around us, and data collection is a part of it.

Ladder Logic Language Languishing

Recently I've been drawn into a struggle between two competing languages. The dispute concerns whether ladder logic is still viable as a control language and the possibility of its replacement by mainstream, serial computer languages.

The ladder concept itself is ancient (well, old, anyway) and can be found on the back of any washing machine made since Hector was a pup. Actually, rumor has it the Germans invented the ladder relay representation in the early 1900s. Ladder logic, as a control language, was first used in conjunction with silicon devices around 1969 at Bedford Associates, the forerunner of Modicon. To support the control language, a hardware platform was devised that had three constituent elements—a dual-ported memory, a logic solver and a general purpose computer.

Early on at Modicon, we used a degenerate form of ladder representation. The great advantage was the language could be understood by any working electrician anywhere in the world. Later, the language was expanded to multinode, and additional functions were added. The rest is history. Ladder logic functionality and PLC adaptability quickly spawned an entire industry.

During the early 1980s we considered "upgrading" PLC technology through use of computer languages. This would have accommodated the young Turks coming out of colleges and universities who only knew serial languages. But everyone said, "Don't change the language. Instead, expand its capabilities to include communications and data base functionality." Would that we had listened! We didn't change the language, but we didn't expand its functionality either. As the applications became larger and more distributed, ladder logic use became onerous for the serious computer or control engineer.

Why was ladder logic so successful? The language was easy to learn. And hard as hell for the professional MIS hack to catch on to. In fact, we purposely did not call the PLC a "computer" so that the control engineer could "do his thing" without becoming enmeshed in the lengthening tentacles of the MIS department.

PLC control, as exemplified by its ladder logic language, was the first rule-based, object-oriented representation by means of computer language.

The technology foundation of ladder logic is not in question.

Users already knew the language, and that language had embedded in it the latest language science. Instead of training users to operate the computer, we trained the computer to work with the users.

The most fundamental language I work with is machine language. For computer design, the basic language is the language of the transistor. The assembler was a fundamental invention that allowed the user to work in a virtual world of instructions that made more sense to humans than machine language.

Languages evolved upwards from assembler to Fortran and C. Then to C++. Languages for today's designer are Forth and LISP. Each step upward allows the programmer to express himself with fewer lines of source code. Use of object-oriented concepts are now fashionable. Rule-based objects and data base representations based upon symbolic notation and object-oriented programming are found in the newest designs.

But wait a minute. Ladder logic uses an object-oriented philosophy in conjunction with rules that solve Boolean equations. For small, non-distributed problems, ladder logic is about four times as productive as normal serial languages. So what's wrong? If ladder logic is so advanced, why is there controversy?

Because ladder logic hasn't gone anywhere since 1969. No vendor of note, nor the academics, has truly expanded the language into non-logic areas. Schools do not teach the language. The computer systems that surround the PLC all speak computerese. To be top-notch, a user has to know both ladder logic and serial languages. Vendors have not supported hardware to allow ladder logic to be embedded with older serial languages so that variables can easily cross over the cultural border. Because the issue is cultural, the technology foundation of ladder is not in question.

The future might see the gradual erosion of ladder logic. And that will mean a loss in terms of software productivity. I will mourn the loss. Replacing ladder logic with older concepts is not the solution. Nor is expanded use of everyone's panacea, function blocks.

To revitalize ladder logic, what I would like to see is symbolic representations in string notation with object-oriented data bases. No system code to write and the source code in a natural language. Some of the techniques to be increasingly aware of are expert systems, rule-based objects, massively parallel software and hardware, fuzzy sets, neuron nets and self-crystallizing concepts in chaos. The future is where we live.

One's a Tool;
The Other's a Solution

A spate of articles has recently appeared in computer journals attacking the functionality of the programmable controller (PLC). In rebuttal, I would submit that PLCs offer a number of advantages lacking in VMEbus and the other open systems these articles advocate for industrial automation. PLC advantages include higher net real-time performance, lower life-cycle cost, broader customer acceptance and a larger selection of systems integrators.

The VMEbus (VERSAmodule Europe) bus standard, used with off-the-shelf boards and modular chassis, can be used to rapidly construct high-performance computer systems. The standard offers advantages such as full 32-bit data path and 32-bit addressing.

To begin, I have to point out that what we're really talking about is two entirely different things. The computer is a tool. The programmable controller is a solution. Programmable controllers, word processors, auto engine electronics and kitchen appliances all have customized "computers" embedded within them that allow them to do one thing well. On the other hand, VMEbus technology can be used to make a PLC or a word processor or anything else you want. But that which is made won't be a standard product. The open system that is created to address the application will be one of a kind.

Customers want standard solutions with good reason. Envision a plant with 300 control boxes. Suppose each is customized for its particular role. Each is almost the same, i.e., totally different. It ends up being like a Chinese restaurant where the menu is so extensive you don't know what to order. If you want steak, it's better just to go to a steak house.

I contend that although VMEbus is useful for repetitious applications such as telecommunications, it is difficult to use in solving control problems. The very versatility that makes VMEbus attractive to me and other computer nerds is a problem for industrial applications. We don't design personalized workstations or word processors. Why should we design a new PLC for each application?

Non-application-specific operating languages are typically used to design tools and only seldom to implement solutions. Tool languages (e.g., LISP, Forth and C++) are not solution languages. Examples of

Control achieved by a random mix of products purchased from the back pages of Byte *magazine makes me nervous.*

solution languages are Ladder, Spreadsheet, WordPerfect, Hyper-Card and industrial workstation facilities from vendors such as Wonderware. Recent attempts to use operating systems as "the" solution have had limited success. UNIX is not a real-time system, although Posix and Realix are promising.

Some of the advantages of VMEbus systems most often mentioned are actually illusory or temporary. FDDI communications available on VMEbus systems will soon be available for PLCs. Any new communication protocol will soon find its way into PLC solutions, once it's proven itself in the computer industry. Sonet, for example, is one of the protocols that will emerge in the next decade. Even E-net electronic mail can be used in a real-time system if the occupancy level is reduced.

Recently I visited a very large science project where the various options available for industrial inputs and outputs (I/O) were the subject of considerable conversation. Interested parties asked each other, should customers be advised to choose I/O on the basis of life-cycle costs or simply on entry costs? To me, the answer is clear. Looked at over the entire life cycle, PLC vender I/O offer-

ings are very cost-effective. What's more important, they are very reliable. Frankly, the prospect of having an elevator, airplane or furnace controlled by a random mix of products purchased from the back pages of *Byte* magazine makes me nervous.

This is not a tirade against VMEbus. It is a tirade against faith in an earth-centered universe. No single tool is the center of an expanding universe of possible solutions. I like VMEbus for the applications for which it is suited. I like PLCs for solving automation problems. That's not to say that PLCs don't suffer from lacks. PLCs can lack flexibility in utilization, lack data analysis capabilities, lack the ability to interchange vendor products and suffer from lack of progress in language enhancements.

But the PLC solves problems. The real issue is software, not hardware. PLC software is "easy-entry" and it works. It is not dependent on a single engineer; it operates in real time, and it is designed specifically for the problem that it addresses. Let the market decide. Why have PLCs become a global business in spite of perceived lacks? Because they work.

The Objects of My Affection

To discuss an alternative approach to factory control software, we again have to talk of real-time computing. A real-time system consists of a controlling system and a controlled system. The controlling system interacts with the environment of which the controlled system is a part. Severe consequences result if the timing and logical correctness of the system are not satisfied.

If we want to look into the future, we need to talk about the latest philosophies. One of the newer approaches to software programming is the object-oriented programming system (OOPS), which has already been very successfully applied in conventional computer systems. Object-oriented technology is both strong and mature in the database and graphics arenas. With OOPS on the real-time horizon, the grail of interchangeable, reusable and updatable code comes tantalizingly into view.

What is OOPS? Think of a Russian egg. You know, the Fabergé eggs-within-the-eggs that the children of the Czar got for Easter. Well, an OOPS uses software eggs, or boxes, that contain within them both the code and the data. The box is called an object. An object can be a single data point or line of code, or it can be made up of complex sets of data and code. The gist of it is that the objects are data-centered, with the code subservient.

There can be many types of objects, including graphical, database and programming units. Objects are both a set of operations and memory "owned" by the object. Conventional languages support a data procedure model in "one long string." With an OOPS, the code and data are treated as blocks, not strings.

Why all the excitement about the OOPS? The use of objects greatly reduces the cost of programming a system. And the technique can be practiced on existing systems. The boxes can be big or small, simple or complex. For dealing with relatively simple records an numerical data, a relational database model may be best. But for symbols and interwoven data, objects work better. This makes objects ideal for user interfaces. One could also imagine ladder logic programs and data used as objects in PLCs.

But factories today do not use OOPSs to control processes. PLCs do not use a distributed objects system in plant control. No sen-

One could imagine ladder logic programs and data used as objects in PLCs.

sor-based objects are in wide use. Why? Because the developers of technology did not try to adapt OOPSs to the needs of real-time systems. However, Digital Equipment Corp. and Hewlett-Packard are working with objects in the context of real-time control systems. We all suspect that the leading PLC suppliers are working on versions of real-time OOPSs.

Let's talk about implementation. The technology should be able to make links from PLC to PC to minicomputer transparent, i.e., the user will not be aware of the techniques used to effect the links. The type of I/O modules involved should not matter either. Interface standards for non-real-time applications that address the issue of object definitions, methods and messaging are due by 1994. Some feel that truly complex systems will not be possible without such standards.

How would such complex systems work? Don't know. However, we can make some guesses. Assume open hardware baskets of I/O. Roaming eggs of data and code wander around and lock onto jobs. Each job is an application of a module in a distributed system. The control workstation can operate with an object anywhere in the system. The workstation is a single point of control. The distributed control system is transparent to the workstation and operates via the objects resident in the modules. Objects must be able to reconcile sensor data with the internal data/code.

Problems still abound. No sensor concepts for objects are technically mature. Yet the problem of reconciling sensor data in large control systems has not been sufficiently explored. Does sensor and embedded data needs to be ergodic? If so, what are the synchronizing issues? Issues in messaging, sensing and operating systems need to be resolved. Language and data representation are not an issue. As usual, culture is the true impediment to utilization.

Objects can integrate and optimize with standard products.

Objects can interface to a complex heterogeneous system.

Objects are independent of location and reusable.

The first vendor to attack and solve the control issues using objects will be the king of silicon hill. To paraphrase Einstein, "thinking is concepts (objects), interconnection and the rationalization of both with the senses." We can expect to see OOPS-based products in 1995 and in wide use by 1999. The PLC of the future is software objects of reusable code mounted on open systems.

Let the Product Be the Judge!

In April, at a Sematech conference in Dallas, Texas, I attended a lecture by Robin Barnes of Bycom Solutions, St. Peters, Missouri. The talk touched on some philosophical considerations relevant to control of the manufacturing process. For one, Barnes suggested that there's more to control than the latest PID algorithm.

She outlined several possible approaches to better product and processes, in particular, predictive control involving modeling and nonlinear relationships. To me, most interesting was consideration of control system configurations that used product feedback as the primary loop, with process feedback serving to stabilize local loops.

Perhaps some definitions are in order. Three basic control and parameter considerations are involved. *Production control* is the management of materials change or modification. *Process parameters* are measurements of process attributes. Last, and unfortunately too often ignored, are *product parameters*, which link product attributes back to the manufacturing steps.

Product feedback gives managers some sense of process quality. Time-to-market and quality are but two possible areas that can be improved by better product feedback. Improvement happens when information relevant to the hierarchical control scheme for each production step, focusing the impact of each process step on the final product. In addition, values such as temperature, flow and force should be directly related to their product values—color, consistency and hardness, for example.

Predictive modeling examines for aspects of a process—those that are controllable and observable; those controllable but not observable; those observable but not controllable; and those unobservable and uncontrollable. Classical process control only deals with linear aspects of the controllable and observable, and usually only by means of feedback. In other words, an error must exist before corrective action can be taken.

Classical control normally involves linear approximations of nonlinear processes, since most processes are nonlinear. This can lead to large product and process errors. (I can see the letters coming in from vendors already. Before going any fur-

It's already too late if an error must exist before action can be taken.

ther, I should say there are products that already do modeling and nonlinear control, but too often only as a patch to the classical system.) Looking at all possible processes, classical systems apply only to a small subset. Even the ubiquitous PID loops use linear algorithms.

Models are better suited for dealing with the unobservable. Models predict future deviations. Wouldn't it be better to make the starting point for process control efforts a model of the true characteristics of the process, one that allows for nonlinear control schemes? The classical approach would thus be a subset of the more inclusive system.

Recent trends of using modeling in real-time control are promising. Modeling proves most useful for processes that have multiple inputs and outputs and nonlinear aspects, that are intolerant of any error, and that must link to product parameters. Modeling and product feedback provide great benefits, especially when simply reacting to deviations is not enough.

Localized process optimization can distort overall production values, leading to poor products. So it's really not overly obvious to emphasize that good products are the whole point. Modeling ensures the entire process is the issue, not some group of loops that's become someone's private obsession.

In general, many of the pre-competitive consortia attending the conference seemed interested in process control. They are also interested in setting standards for open systems. Most, however, are only interested in the standard concepts of linear loop control. The majority of the attendees talked about wanting standards, modularity and open systems. But, perhaps unfortunately, too little was said about that which all our efforts are aimed: better products.

Issues of Real Time

Philosophers and would-be philosophers look to the meanings of words—sometimes even technical terms. We've talked about real time before. It refers to tasks or functions executed so fast that their feedback guides task completion. Real-time processing is fast, robust and predictable.

Real-time control models a process to act upon it. Data coherency from sensors and presentation of the values via the effecters is important. The sampling and hold system must fit within the performance envelope of the system designer's requirements.

A little appreciation for the theory is in order. Sampling theory says we can reproduce analog wave forms by sample and hold, if appropriate conditions are met. Each baud must contain at least one valid sample of the amplitude. Generally, engineers use the rule of thumb that 4 samples per cycle (or 2 samples per baud) will adequately reproduce the wave form without much processing.

Simply oversampling—with the idea that more is better—seldom enhances accuracy. The benefits of frugality can be seen in the compact disk (CD) player. Clearly, CDs provide the best reproduction of sound available. It's only the single transformation (in DDD disks) of the analog wave form to digital that can introduce error. Digital trans-

mission prevails throughout the rest of the reproduction system.

Several factors influence good analog-to-digital conversion, which is essential to any measuring. The most obvious is sample rate. It must be fast enough to satisfy the theoretical requirements. Less obvious is the need for predictability of the sample rate. Otherwise, unwanted modulation products emerge.

Another requirement is to establish digital conversion accuracy, which can vary from 8 to 21 bits, with the typical range from 12 to 16 bits. Ten bits has an unknown value of one part in a thousand, 20 bits is accurate to one part in a million. To put that in perspective, consider that temperatures here in New Hampshire range from 20 degrees F below zero to 110 degrees above: a range of 130 degrees. The body is sensitive to about a degree, or one part in a hundred. Eight bits of conver-

SPEED ZONE AHEAD

Analog-to-digital conversion is the essence of quantification.

sion accuracy should satisfy home thermostat requirements for anyone except my wife.

Conversion is meaningless unless the held sample is representative of the "instantaneous" value being measured. If the digital conversion takes significant time and the signal changes during that time, errors result. The sampling aperture must be such that the capture time is within the error budget of the process. This is an oft-neglected, major contributor to error.

In summary, a designer must consider the aperture, sample rate, sample coherency and number of bits. What has all this to do with computerized process control? Plenty. In a classic small, real-time system, the designer must consider similar elements. Quick response ensures that the system is sampled consistent with theory. Interrupts must be grouped and prioritized so that the aperture of capture is within the change limits of the desired system. Software error budgeting must bear in mind the number of events, the change rate of sensed data and the service for each event. Real-time operating systems take all this into account. Too often, novices assume any operating system can be real time. Disasters result.

The PLC, however, has a scan time of about 5 milliseconds inde-pendent of change. The PLC operating system ensures consistency of product, independent of stimulation. Predictably, many military computers use real-time operating systems or modified common operating systems. To use a modified, PC-based operating system, you must know—*a priori*—the system requirements.

In any industrial application, programming types, operating systems and I/O structure must be defined. Real-time considerations should be part of each systems analysis. We, in the business, tend to become enamored with buzzwords. Distributed systems have better real-time performance locally, but penalize overall system performance. Centralized systems penalize local-node performance with attendant superiority overall. Classical string languages and operating systems have excellent interfaces and database performance with marginal consistence control aspects. Using appropriate tools heterogeneously is best for now.

Silicon in Control

Recently this columnist attended a workshop on semiconductor manufacturing held in San Antonio, Texas, home of the Alamo and the Rough Riders. Next to the hotel was an Imax theater showing a special Alamo film and *Jurassic Park*, three stories high. Guess which one your token nerd saw.

This workshop—one in a series looking at manufacturing problems in the semiconductor industry—focused specially on wafer fabrication problems.

Anyway, I had to listen to that most common of all laments, "My control problems are special." Years of experience have taught me that control problems are similar across many industries. The secret is to locate the true differences that make the problems special. Issues involving materials handling, software, yield (quality) and diagnostics are common across the control spectrum.

What's special about the semiconductor industry can be summarized as follows:

- fierce competitive environment;
- fast rate of product change;
- tight, complex tolerances; and
- the importance of the industry to the nation.

It pleased me to see there was considerable interest in learning how to better maintain control of processes battered by unforeseen events. This led to many hallway discussions about chaos and emergent systems, subjects close to my heart.

A wafer production line must meet process specifications in the face of the extreme swings of multiple variables. Any control system needs to be insensitive to stimulation. But in wafer manufacturing, this requirement is crucial. Quick time-to-market and high throughput require use of the latest in modeling and feed forward control.

Imagine a spread sheet of control solutions. The horizontal axis presents an increasing level of technical complexity, from direct feedback control involving a few sensors and effectors to plantwide kilosensor/effector systems.

In the vertical axis, we consider the sophistication of control type—of solution, not problem—from direct one-to-one control-to-sensor fusion to full-model control with dead reckoning. In other

The control system of the future will be a model of the process running faster than real time.

words, a spreadsheet of problem "size" vs. the solution technology.

For example, early PLCs controlled via direct binary feedback based around "limit switches." PLCs today are used in large kiloplant systems.

Personal and minicomputers were used in the early days for intelligent control, but lacked the power to handle fast, large, multipoint systems. Combining the PC and PLC is what's done today.

Model-based control will be next. Modeling is best explained by an example. A wrist watch does not tell time. It models time. It is, at best, an approximation of real-world time.

Like a submarine, the watch navigates by dead reckoning, you reset it once a month and that's good enough. The submarine uses a model of the currents, and awareness of its approximate direction and speeds to "guesstimate" its current position. Course correction is done by surfacing and using a sextant.

The control system of the future will be a model of the process running faster than real time. Feedback from the modeled process will impact the real process. At appropriate intervals, the process model will be re calibrated for further use.

Parameters for wafer fab, for example, can be determined using feedback based on predicted performance—a control scheme resident in the database, with update from sensors. Sensor busses need to accommodate large quantities of dumb sensors and/or small quantities of smart sensors.

Updating of the real processes' I/O data will be done asynchronously from the model. Process parameters will feed forward into the real manufacturing process from the model runs.

Wafer fabricators can only use, not drive, existing control offerings. Many fabricators are beginning to concentrate on product, and have forgotten about reinventing the control industry. This growing maturity will lead to a robust, agile process. Working with the technology providers will lead to a broad spectrum of control techniques. The industry is off and running. Congratulations, and keep on trucking.

Rise and Fall and Rise

Once again this spring I made my annual pilgrimage to the IPC show, which used to be the International Programmable Controller show. I can still remember the very first one, at least 25 years ago, with only half-a-dozen exhibitors in the cellar of the EDS building. Over the next 20 years, the show grew until, each year, for the better part of a week, it occupied a very substantial portion of Detroit's Cobo Hall. The show's growth was a reflection of the fact that an entire industry had grown up around the programmable logic controller.

The programmable logic controller (PLC) is to factory automation what the PC is to office automation. It is a microprocessor-based device that stores instructions in programmable memory and implements logic, sequence and timing functions based on inputs from limit switches, push buttons, thermocouples and other devices. Introduced in the automotive industry to replace relay boards, today even the smallest PLCs may be equipped with serial communication and analog control capabilities for loop control of flow, level, temperature, pressure and other variables.

Now, 25 years later, the PLC industry has

gone though several waves of consolidation. Although still in wide use, the PLC is more of a commodity item and there are fewer programmable controller manufacturers than there were several years ago. More and more, the PLC is to factory automation what the PC is to office automation.

These facts, too, have changed the IPC show, which isn't quite what it used to be. In fact, next year it will be called the International Automotive Manufacturing show. The pace of technology change is now such that one cannot only see something new appear on the face of the earth, one can also, in that same lifetime, see it disappear.

While at the conference, I sat on a panel of experts discussing the current status of the PLC. As the pundits argued endlessly about various standards, flow charting, Microsoft Windows and laptops, global competition and technological turbulence, we were eating our lunch. It seemed to me that the panel—and the attendees—should be concerned with the future in-

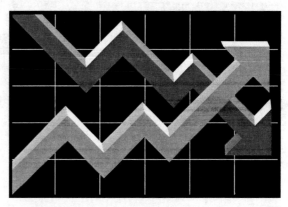

Look for a future not of computers, but of appliances.

stead of the past, and I said so. Actually, what I said was that Detroit seemed mired in technology's third world. I gotta do something about my temper.

Now, a more appropriate subject would be something like the impact and direction of computer chips. Processor performance improves by a factor of 10 every five years. Or a factor of 100 over the next decade. Right now, RISC computer chips are running at about 150 megahertz. By 2001, they will be running at 1.5 billion hertz. And in 2006, 15 billion hertz is predicted.

Object-oriented software development promises that in the future fewer lines of code will be needed for equivalent functionality. But to enact object orientation, a corresponding increase in computing power and compiler know-how is required. New, powerful chips will be needed because of the new kinds of software.

Applications of the new power sound like things out of a world of science fiction: speech translators, PLC web platforms, disbursed databases, sentient behavior and transparent use of control. In general, software models of "real" systems will be used to monitor and control those systems.

One company I'm associated with, Flavors Technology, builds hardware and software for rule-based agents that have emergent properties. Initially, we had "only" 680xx chips available for use. Up to 128 of these chips were used to get 16,000 agents all running in parallel at 60 hertz. The boards were 22 inches long. Now we do the same with standard VME cards. Our Asian customers want even more power—up to a million agents all resolved in the same 16 milliseconds. This will be possible over the next decade.

Applications, in general, will move off the desktop and into real time. Word processing has gotten about as good as it needs to get. After all, we can only type so fast. But planes, trains and automobiles, scheduling and data processing can all make use of the possibilities inherent in the convergence of the Worldwide Web, expert systems, LAN legacy support and emergent capabilities.

We need to gain a greater understanding of the new software technologies. We need compiler experts in parallelism and agents. We need semiconductor resolution at the sub-micron level—at least down to quantum effects. We need chips with memory bandwidth. What we need is an I/O firehose to make the CPU chip work. Look for a future, not of computers, but of appliances.

Floor Wars

Manufacturing *Systems* is dedicated to the premise that manufacturing managers must make use of information technology to improve productivity. And, indeed, enterprise planning and manufacturing execution systems have materially changed how the production environment is managed. However, process control must still be accomplished on the plant floor. Today, most production lines are controlled by programmable controllers (PLCs) and managed by the ubiquitous personal computer.

It seems that about once every seven years, an argument breaks out concerning time-share vs. local computing, "new" programming languages and the floor war over computers and PLCs. Lately, major segments of the trade press have decided that all PLCs can be replaced with PCs. This is not going to happen, but changes are occurring.

The PC is making inroads in real-time control, most especially for "soft" (non-deterministic), real-time control problems. I estimate that PC penetration of traditional real-time PLC business will be four percent by the end of the year.

Everyone loves a fast-growing market. Any market

that goes from zero (with apologies to some vendors) to $500 million will draw all kinds of attention. Heretofore, the process-control business has been stealthy, and well hidden from the *Forbes* and *Fortunes* of the world. But now it has the light of publicity upon it.

Some history is in order. PLCs were invented to offset some of the defects in standard computers. The PLC:

- has no on/off switch;
- has no rotating components;
- operates in several physical environments;
- can be programmed by the union trades;
- has stability as a product;
- features object orientation in the form of relay ladder logic;
- is a hard, real-time operating system; and
- has insides made of PC hardware

On the other hand, the computer/PC has:

- excellent communications;
- superb graphics display;
- database capability;
- infinite memory;
- PLC hardware for the co-processor; and
- excellent soft, real-time capability

A *hybrid PLC/PC approach to process control is today's best solution.*

Today's engineers arrive with C++ programming experience—not real-time, ladder logic skills. This naturally leads to greater use of the computer as a controller. One consequence of this trend is that it's more difficult to make changes. The suggestion that Murphy's Law no longer holds true bothers me. We, as engineers, are wrong to assume that we can design and deliver a system that requires no follow-on changes. We call this "no-change maintenance" in the software business.

However, the process engineer or operator may know better. I have never known a control sequence that remained unmodified from its inception. Even automobile engine chips have an aftermarket for improving some parameter.

I have always assumed that technological progress and change was good—until I talked to the user community. Many of the complaints were about chips, peripherals and software becoming unavailable in as little as 18 months after introduction. Both the PC and PLC vendors fall into the trap of quick-time. It used to be that 20 years was the half-life of processing assets. Now the beanies use a five-year life cycle. But please, not 18 months.

Recently, I was a speaker at a PLC vs. PC conference in Australia.

The majority of the panel sequestered for the wrap-up agreed that the business solution is the best solution. Consider the best interests of the process to be controlled, not the allure of adventure and risk. In general, process management and the human interface should be handled with PC technology. Hard, real-time I/O should be handled by PLCs.

Further, MicroSoft Windows NT will soon be market-dominant. The I/O of the PLC vendors seems best for sensor and effector applications. The PLC does hard sensing in real time. It is a physical asset controller, adaptable to the service and operations aspects of manufacturing.

On the other hand, the PC is here to stay, and with increasing market share. This suggests a low-risk, hybrid approach to control issues. Don't fall in love. Today's solution is a Windows NT topside with PLC co-processors. Let the future evolve by itself. To a hammer, everything is a nail.

Part II —
Industrial Computing:
The Camel Has the Tent!

Widespread uses for PCs in industry are still as operator or HMI (human-machine) interfaces, especially for PLC-based control systems. They serve well for data acquisition, programming and communicating with area and plant networks. Tremendous advances in support software have led to timely QC/QA programs, CAD/CAM links, diagnostics and preventive maintenance, simulations and on-line modeling.

Laptops and now palmtops are routinely used for accessing control loops via phone lines, Internet or wireless and are proving invaluable for minimizing geographic disruptions.

Articles Comprising Part II:

DEC Chips in With Alpha • Faster Than Real Time

Super Computers in Command: A Look to the Future

Heat, Light and Hardware

Intelligent Materials Part I: The Dawn

Icons, Glyphs and Strings • Seeking Better Bandwidth

"The measure of success is not whether or not you have a tough problem to deal with, but whether it's the same problem you had last year"

—*John Foster Dulles*

DEC Chips in With Alpha

Recently at DECworld in Boston, Digital Equipment Corp. (DEC), Maynard, Mass., demonstrated its reduced instruction set computing (RISC) design for a chip-embedded computer architecture. DEC was able to show that the design, called Alpha, clearly outperforms the latest available workstations competition. But your first question might be, why has Digital gone into the silicon chip business?

The answer is, because DEC knows proprietary systems are a vanishing breed. Digital will sell Alpha at all levels of integration—chip, board and system—to other companies and to OEMs. DEC's strategy is all about very powerful workstations networked into specialized servers. As Richard L. Sites said on the subject of Alpha in the August issue of *Byte* Magazine, "The microprocessor has had as much impact on the economics of computing as it has had on the performance of computers. As a result, any new CPU [Central Processing Unit] is likely to be conceived as a single-chip design." Alpha both defines the workstation and is DEC's entré into the open systems world.

Alpha architecture, which is compatible with existing complex instruction set computers (CISC) and with DEC's VAX architecture, is getting long in the tooth.

Continuing to support the VMS legacy, while looking ahead to RISC in Windows NT and UNIX, reassures us old guys that we're still loved and positions DEC in the workstation marketplace of the future.

Technically, the chip is a whiz. The demonstrated clock speed at DEC world was 150 megahertz (MHz). Digital says it's the world's fastest microprocessor. Before very long, speeds of 200 MHz, and eventually 400 MHz, will be available. Very few applications today approach the limits of 32-bit addressing, so Alpha's 64-bit processing capabilities back up DEC's statement that this is an architecture for the next 25 years.

On the other hand, Alpha may not be the ideal chip for embedded and real-time applications. For example, blazing speed in floating-point manipulation is not a prerequisite for real-time performance. Higher priority is the representation, in integer fashion, of the large numbers generated by process control or other applications. Although floating-point overflow is handled well by Alpha, a rational approach to integer overflow and recovery is needed.

Alpha is important even if it isn't ideal for real-time and embedded applications.

The chip's caches are adequate. But because larger caches are needed for today's control systems, 8k bytes ain't enough for the on-chip data and instruction caches. My guys would like to see 64k for the caches. Because we "don't need" floating point, we'd prefer to have larger caches and treat floating point as an off-chip problem.

Alpha is important even if it isn't ideal for real-time and embedded applications. After all, real-time use constitutes only 10 to 15 percent of the market. Overall, technically, the chip meets the important benchmarks. But simply meeting benchmarks is not the name of game here.

DEC's change in focus from computer systems to a chip-embedded architecture is serious business. Even for a multibillion dollar company, competing in the processor chip market can tie up a lot of resources. Moving away from what has historically been the company's focus can be risky. But if the market accepts the chip, it will be a big win for the guys in Maynard.

Digital faces both technology and market challenge. Chip technology has so far gained by about a factor of ten every five years. If that trend holds, in 1997 Alpha would run at 1,000 to 2,000 MHz. In 25 years, Alpha's power would have to be 100,000 times today's performance.

In retail sales, the three most important factors are location, location, location. But for the semiconductor business it's timing, timing, timing. Even if you introduce a product twice as good as anyone else's you only have a little time to exploit your advantage. The competition will immediately design to technologically "leapfrog," based on what you've done, until they archive parity. And if there's a delay between announcement of a prototype and manufacture of production quantities, the "exploitation period" is commensurably diminished.

DEC has entered a tough race against experienced horses. Alpha is a good way of getting out of the gate. But is it too little, too late? I agree that it's an architecture with a 25-year life span. But 10 years have already gone by.

Perhaps what alpha does for DEC is similar to what sports cars like the Viper or the Corvette do for an automotive company. It demonstrates technical prowess, helps sell product and gives engineers something to sink their teeth into. DEC's efforts are to be supported. What's good for the computer industry is good for all of us. Hurray for Digital Equipment Corp. and best of luck.

Faster Than Real Time

Most of us think of simulation as "merely" objective representation of things or processes. And they are. But bear in mind that language is also a kind of representation. "Natural language"—English, German, Chinese—is a map of the embedded belief of a culture. It contains one element of a game aimed at attaining some goal, be it getting work done or sharing a meal.

Computer languages such as Ladder, Drum and Paracell are all representations that reflect a particular "brand" of problem-solving. The same could be said of Lotus 1-2-3, HyperCard or Windows. Natural language, computer language and even commercial computer programs are all representations or simulations. The form of the representation is determined by its use.

Another type of simulation, of interest to manufacturers and engineers, allows modeling of chemical and other kinds of processes. Most simulations do not run on a real-time basis, nor do they include direct sensory input from the process in question. A simulation tool frees engineers to tinker in a world of make-believe, to examine relationships between ingredients and play "what if" games. It allows them to methodically identify strategies for process optimization.

Computer-Aided Design (CAD) is simulation in the discrete manufacturing realm. Enhanced and virtual representations of the mechanical engineering world enable design optimization and concurrent engineering. Models can be used to examine relationships between objects for interference checking and allow structural, thermal and other type analysis.

Narrow definitions for computer-aided design and simulation limit its application. A simulation must be a model of more than "just" a product or a process. Recently I visited Lockheed, where there is considerable interest in "space-age" composite materials. Controlling a composite's characteristics during its production can be difficult. And to get a handle on it, Lockheed has to be involved from the well head to the warranty expiration. It wants to be able to "simulate" the entire life cycle of the composite material.

The design environment and associated system of today's simulation technologies are typically brittle, especially near the edge of the performance envelope of the software. Users either have to be experts in manipulating the platform or they have to bring in some-

Obsolete, OOPS, Rule-Based

Natural language, computer language and commercial computer programs are representations or simulations. The form of the representation is determined by its use.

one who is. But there is growing awareness that those who know the products and processes to be modeled must themselves be capable of manipulating the simulation.

Today there are three kinds of software; obsolete, OOPS (object-oriented programming systems) and rule-based systems. Two of the three hold within themselves the promise of user enablement—of increased productivity though advanced technology for more flexible manufacturing. OOPS and rule-based systems capabilities will be enhanced by direct coupling of the simulation to the "real" world. Real-time modeling is what will take process optimization to the next level.

Real-time computing is fast, robust and predictable. And when modeling, the big question is always just how fast, how robust and predictable do you want to be. Much like with a compact disk player for the home CD, the process must be over-sampled to make sure nothing is missed. The model must therefore be "faster than real time," and be able to anticipate how the modeled process will react, so that comparisons and projections are ready prior to an event's occurrence.

We know a incandescent light bulb lasts 1000 hours. So it should be replaced after 980 hours. In the same way, we should be able to predict bearing failures in power turbines used to meet peak demand in power utilities. Predictive diagnostics are a powerful tool. But it's difficult to build a computer system capable of responding to what's going to happen tomorrow. To be real time, modeling systems must have the means to foresee developments.

A couple of other quick points. The mathematics used in most simulation packages assume linearity, for load as well as structure. This is a limitation. It may be that the use of standard operating systems such as DOS or UNIX hold back user enablement. Suppliers should examine the usefulness of a multi-user, object-oriented data base with n-dimensional representation.

Simulation and modeling technology today is at that level attained by aviation technology just after the Wright brothers arrived at Kitty Hawk. Someday, simulation capabilities will be comparable to the accomplishments of Boeing in the aerospace industry. Correctly used, we'll be able to see into the future.

Supercomputers in Command: A Look at the Future

In the February issue we discussed the 1993 Santa Fe Chaos conference. Reader response to that column was quite high, and we'll continue to keep an eye on chaos theory and how it relates to industrial control. Careful study of the characteristics of chaos and control can help solve those sticky applications whose variables inhabit that nebulous region that lies somewhere between linear relationships and truly random events.

In the mid-'80s, I was doing some work on very-high-speed, rule-based systems for an outfit located inside the Washington beltway. That's not unusual—radical technologies are the life blood of the defense establishment. Some associates, however, learned of the government work and suggested that these "chaotic" technologies be used to solve problems in discrete manufacturing. Here, at Flavors Technology, we were incredulous. Manufacturers—we thought —were only interested in incremental improvement. But times had changed.

It's possible to see why the change occurred. In the days of wood-en ships and iron men, all control was done with relays. As electromechanical control systems become more complex, the rows of cabinets containing relays reached lengths of 50 feet or more. The invention of the programmable controller (PLC) was the means to shrink those cabinets back down to five feet.

But solving that cabinet problem allowed developers to further increase control system complexity. The rows of cabinets began to lengthen again, this time filled not with relays, but with PLCs and computers.

We're convinced that supercomputing power and the science of emergent computation will be the means to shrink those cabinets once again. We sold the business of supplying the defense guys and are concentrating our efforts in the industrial arena.

The computing model in a plant usually has five distinct layers, from the information processing centers to the work floor. We want to reduce those five layers to three: the processing center, the plant host and the super controller.

We were incredulous. We thought manufacturers were only interested in incremental improvement. But times had changed.

We knew that use of massively parallel, modern computer architectures would allow adequate power to be brought to bear. And in a single stroke, we could shrink the node count from 1,000 to 10. But you'd think a plant configured like that would have to have software of almost bone-crushing complexity. We had to convince our champions at General Motors by using the ultimate argument—the Japanese had bought a unit. GM was convinced.

Production hardware and software were delivered in 1991. A high-level rule-based language was used (English) and a decent graphical user interface was installed. Using chaos/complexity technology and supercomputer architecture reduced hardware and software by greater than an order of magnitude.

A printout of the programming that had been used for the application was a stack of paper 10½ inches high. The first cut by GM in programming, based on our system, reduced the stack by a factor of 13. Since then, we've continued to improve, so that the program is now only several pages long.

How is this done? The architecture is software-centered. We use a multiple-instruction, single-data model, with direct connection to the system's minicomputers and PLCs. One memory is accessed by many thousands of rule-based agents. Each agent has only about three rules. The agents support most mathematically describable functions. PID, fuzzy logic and neural networks all work well. Inputs and outputs are directly imaged into the memory, with from 100 to 15,000 agents installed over time. With a simple set of rules boiling up from the bottom, the control system evolves from within the computer archtecture.

We have, in effect, a real-time game of factory CIM that is fun to play. We, and others, are also tackling problems in power nets, composite materials, semiconductor design, economics and dynamic scheduling—all in real time.

The software is seductive. The power is intoxicating and the technology is real. It is exciting to be a technologist in these times. Come join us.

Heat, Light and Hardware

When I started in this line of work, we were still in the era of wooden ships and iron men. In those days of yesterdecade, we did the algorithms with the hardware topology. Now we make magic with the software. But it's hardware that houses those software systems.

There are three basic configurations. First the classic single processor such as in a personal computer (PC). Second, node network of both heterogeneous and homogeneous single processors, as widely used in business and industry, e.g., local area networks and client/server computing. And third, the most recent innovation, massively parallel processors (MPP). These are normally homogeneous and installed in a single box. From a designer's standpoint, all three hardware configurations are related and differ only in details.

The PC is but a node in a system. A network is several nodes put together on an ad hoc basis. In MPP, the designer formally configures the assembly of nodes for special purposes. The classic super computer is a small group of high performance vector processors in a box. Multiple nodes are then grouped for more power. Eventually we end up with a massively parallel processor.

Obviously, even computers gotta have the hardware components of processor, memory and I/O channel. These elements can be configured in the any of four basic forms. The single instruction, single-data architecture is used in PCs and workstations. The single-instruction, multiple-data architecture was used in early MPP finite element analysis applications. The multiple-instruction, multiple-data approach is used for distributed control systems in process industries. Least common is multiple-instruction, single-data used for real-time control in programmable controllers, as well as Flavors Technology's PIM system.

The I/O is usually not considered as part of the initial architecture. This is changing, however. Computer scientists understand now that a "philosopher computer" solipsistically performing logic functions has only limited application in the real world. The basic architecture should include all three: processor, memory and I/O.

The fundamental constraints in system design are

It takes two years to nail down a concept and two more before you ship.

heat and light. Heat dissipation is the one of the most frustrating because we always feel we should be able to overcome it with flow and dissipation mechanisms or by using less power per calculate function. We don't feel so bad if we can't overcome constraints that are a function of the speed of light. We can skirt around those constraints by using parallel processors and reducing run length. This has a ripple effect, as semiconductor chip become more complex in order to work in large groups, and special software is written for the chips.

When you design a computer you don't manage by setting a goal. You start by managing the available resources. It's not in our interests as engineers to design machines that can't be built. On the other hand, because silicon performance in memory and processor power improves by a factor of 10 every five years, a designer must plan for chips that don't exist yet.

Design for manufacturing must also be considered. It's important to guess right. What chips are available? And when? What board sizes? How many layers? What software is there? What will new software cost? And most of all, who will be the user?

Recently, we designed a machine for export that had to meet the Department of Commerce specifications. Meeting those regulations had to be our top consideration. Other considerations might be, what kind of data base is needed? Is the system for real-time control, or is it for an MIS department? We don't design simply for our own edification, but for an application or use.

The team approach is strongly recommended. We now know that skunk works work, solving problems in half the time and at one-fifth the cost. It takes about two years to nail down a concept design and about another two years till you have a significant customer shipment. Note that these estimates are for a new architecture, not simply building a "special," or updating a current design.

In the future we can expect to see memories in the Kgigabit range, gigabit solid-state memories, 100 MIPS available for desktop computers. Swarming software and autonomous agents will alleviate the need to generate huge amounts of software. Ever-accelerating power and intelligence will increase over the next five years to astronomical levels, possibly beyond our ability to apply it. It's where we're going.

Intelligent Materials Part I: The Dawn

One reader writes: I was talking to my old buddy Rusty the other day. Actually, that's not his name, but he's a metallurgist with red hair, so of course that's what I call him. Rusty's all right, even though he likes to glorify the role of metallurgy in the history of civilization (he likes to call metallurgy the second oldest profession). I asked what was new in the metals processing world.

"Chaos," he said.

"What about chaos?" I reply. Rusty knows I go on the occasional diatribe about chaos theory and its all-encompassing uses. So I'm a little suspicious here that he's trying to give me the business.

"I've been thinking about hypersensitivity, unpredictable behavior and complexity, which you say are the characteristics of chaotic systems," he said.

Now I begin to warm up to the conversation. I've been preaching this stuff to Rusty for some time and he'd never shown any real interest before. Perhaps there is hope for him yet, I'm thinking. "Go on, go on," I urge.

"Seems to me that if there are any words that describe most metals processing activ-

ities, they would be complexity, hypersensitivity and unpredictable behavior. I wonder if applying chaotic ideas would give us a handle on how to control these processes."

I have to suppress a chuckle at this point. Ever notice that people will accept new ideas more easily if they believe they thought them up themselves?

"Really?" I say, leading him on. "How are you going to predict processes that are unpredictable? And what are you talking about, specifically? If materials processes were unpredictable, how come we can make all the different things that we do?"

(This way I'll see if he's really caught on or if he's just making idle chatter.)

"To answer your last question first," Rusty began, "most manufacturing processes are developed from laboratory experiments into full-scale industrial, repeatable, controlled processes over long periods of time, with a lot of

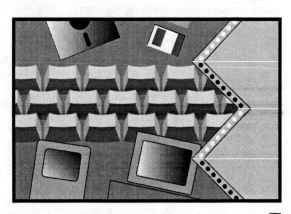

Rusty sometimes treats me as if I live on a farm in New Hampshire.

guesswork and human experience thrown in. Once we guess the right set of points, we program the PID to hold everything in place and then make stuff. Works great if the stuff you're making is a bazillion widgets. But it doesn't address the problem of cutting down the cost of developing new processes and materials, or how to do it quickly and efficiently. Ever heard of 'agile manufacturing' or 'flexible manufacturing'?"

Rusty sometimes treats me as if I live on a farm in New Hampshire and never get to look at the latest issues of Manufacturing Vogue. Okay, I figure, I'll play his silly game. "So chaos is going to help control unpredictable processes?" I ask. "How?"

"It goes something like this. We recognize that the refinement, solidification, deformation and joining processes are inherently hypersensitive to external conditions, which means just about everything in and around the process. So we model the process, not with linear finite-element models, but with independent agents, object-style modeling, and we use the model to determine where and how the processes behave in a chaotic or complex fashion."

"What does that buy you?" I ask. After all, Rusty is still only repeating just what I've been telling him for some time now.

"By itself, that gives you an understanding of the process's 'dynamic system behavior,' which is no small thing. But we want to do more than just map the state of the process; we want to move the process into states that we determine will produce the material, or formed product, that we want. So we link the simulation, in real time, to the sensing and control connections. Then we 'predict' the next state of the process based on the current state, which we can measure in real time, and the simulated new state. If the program doesn't like the predicted outcome, it makes changes to the controlling output, which it uses the simulation to check, and pushes the system toward that state. Simple."

"Sounds like you need a really fast machine to do all that simulating, sensing and controlling," I comment. "I wonder if such a machine exists?"

Then Rusty smiles a smile just like the smile of that goofy Warner Brothers cartoon character who says, "Eh, could be."

Then Rusty says, "Eh, could be."

—Alan Campagna, President
Theta Systems, Woburn, Mass

Icons, Glyphs and Strings

Recently, I decided to administer a functional literacy test. The test was simple: take a favorite Windows application and tell me what all the icons mean. The icons are the droppings located at the top and sides of the Window display. In a surprise quiz, no one I tested could fully explain the command icons. Fascinating.

Perhaps a little history is in order. Way back, in the beginning, we depicted "things" on cave walls. The drawings represented animals and tools. Not much use for action verbs, but good enough to function as cue cards for a story teller. Verbiage began to take on talk-show characteristics.

From these original "icons," humans derived the more complicated icon-based written languages. Early Egyptian and Chinese writing converted icons into hieroglyphics. Hieroglyphics are, by definition, hard to read, perhaps purposely so, since many of these systems of abstract images were meant to protect the prerogatives of the priesthoods of yore. Since the glyphs are not directly representative of some thing, translation is impossible, without a grounding in the culture of origin.

The temptation to use intuitively understandable representations as a substitute for natural language is strong. Icons—as opposed to glyphs—are easy-entry concepts. A human can assimilate several representative icons and immediately use them in simple applications. Computer icons first began appearing in the early Apple Macintosh. A wastebasket was for disposing items. A folder was for storing stuff in groups. On the caves of France, a drawing of a deer was a deer.

Examples of glyphs abound: the oil warning light on your car, most of the new Windows symbology. Hieroglyphics are impossible to read without some insight into the reasoning of the glyph designer. Only the designer knows for sure.

Moving along. Some cultures eventually decided to make use of a fixed number of symbols and string them together as a means of communication. Although the symbols are abstract, they're limited in complexity and number. We call this set of symbols—this acoustic representation of language—an alphabet.

Why can't we learn from the lessons of history when designing computer-interactive displays?

Early string languages were batch-compiled. With most Romance languages, you had to wait until the end of the string expression and then figure it out. English and English like languages have the advantage of incremental compilation. We can construct the meaning as the sentence is received. The English language is widely used because of its intrinsic structure in which incrementally compiled strings can express things, actions and ideas. As usual, form follows function. English is not an easy-entry language, but once entered, can enrich all.

It's important to remember that written language is a system of interrelated symbols, not stand-alone hieroglyphics. Scholastic achievement scores have not materially changed since 1945, but the cost of maintaining those scores has quadrupled in real dollars. Why, you ask? Well, for one thing, the start of the cost increase coincided with the advent of "flash cards" for teaching reading. Flash cards treat each word as a glyph and not as a set of symbols. A child learns to read a limited number of symbols (words) very quickly. But as new words are encountered, the fledgling reader is stumped.

Why are we repeating the same mistake in the design of computer interactive displays? We like icons because they represent real objects. And there are not too many of them. We come to believe that since some are good, more is better. Glyphs derive from icons, and thus, abstract symbolizes arise. History repeats itself. For example, there are glyphs in Windows and Windows-like applications, and the libraries for CAD/CAM application packages. Even the new standards for communication use the library concepts.

What can we do? As users and buyers of packages, we should insist that we, the users, can converse with the computer, not just choose tasks. We want to define the tasks ourselves. Consider easy entry versus effectiveness. Both ways have uses, but we should not simply accept the status quo. Control displays need to have large, single-message displays, in the language of the operator. Maintenance people need clear instruction to service machines they may never have seen before. Use the history of language as a guide, and let's not repeat the same mistakes.

Seeking Better Bandwidth

Someone asked me to check the possibility of managing—or directory controlling—industrial process over the World Wide Web. There are several problems therewith. Security, response time and bandwidth to begin. If we are to "tele-operate," it's always been assumed we need fiber optic connections throughout the country. The implied infrastructure capital costs are immense.

As for bandwidth, every factory and warehouse will need T1 (1.5 megabits per second) performance to each drop or service location. And, the protocol would have to be interactive in real time, with existing interconnect systems preferred.

It turns out that telecommunications companies have been facing the same problem, and that ADSL (Asymmetric Digital Subscriber Line) is one possible answer. This revolutionary concept allows high-speed interactive bandwidth in existing installations of copper loops such as are found in today's factories. And at low cost.

Use of ADSL could leapfrog that of modems and ISDN (Integrated Service Digital Network)—another possible means of high capacity data exchange. Modems are "stuck" at 33.6 kilobits per second,

before data compression. Understand that Data compression is not the same as raw bandwidth. Data compression techniques are able to get a virtual bandwidth four times the raw bandwidth. This discussion will stick to the bandwidth story and leave data compression for another time.

ADSL's reported ability to approximate for T1 lines (about 6 megabits) over the installed copper twisted-pair met with initial skepticism from this reporter. But skepticism withered under testing and demo installations. ADSL is reliable, robust and cost effective. It's blowing people's minds.

ANSI and other standards groups began work on ADSL in 1992. The first issuance considered a capacity of 6.2 megabits per second over 12,000 feet of 24-gauge twisted pair. Whew! Longer loops using "only" single T1 (1.544 megabits per second) can be used to 18,000 feet. The "modems," or processing units at the drops, require 1.5 million transistors.

Trial deliveries over copper have been impressive. "We ran ISDN, for TV channels, and POTS (Plain Old Telephone Service)

The possibility of controlling processes over the Internet.

simultaneously without a glitch," said one participant. ADSL with four T1-capacity has also been heavily tested. Motorola, Analog Devices and others are producing the hardware.

How does it work? ADSL's impressive transmission performance over long copper loops is achieved using advanced digital signal processing techniques implemented in VLSI (Very Large Scale Integration) technology.

The concept is that communication is asynchronous. A page of text is one to two kilobits. Compressed graphics is 10 to 100 kilobits. The control needs are only a fraction of the downlink requirements. One percent to 10 percent is normally the ratio mentioned of control to downlink. The "low speed" capacity of T1 uses an unlink control capacity of 64 kilobits per second. The downlink capacity of 6 megabits per second (four T1 "lines") uses 640 kilobits per second of control capacity. The basic modulation scheme uses discrete multiline modulation (DMT).

Present ADSL is expensive, about $2,000 per drop. The cost will soon be only about $500. And we can expect to see modem-like pricing by the turn of the century. Typical T1 costs are $200 per hour, while ADSL is projected at $15 for the same service.

What are the problems? The techies have "wired" every home and factory with "fiber" performance using the phone company twisted pair. But no single vendor has all the bits and pieces for the phone companies, and no one has looked at systems for manufacturing. Vendors are typically small and thinly financed. Fiber-optic and coaxial cable are coming on strong worldwide. Cultural and regulatory resistance can be expected to this or any innovative technology.

ADSL uses untreated twisted pair loops to yield video and T1 performance. ADSL, or similar concepts, will change the installed copper loop capacity. Our factories and plants are wired for it today. It will be a quantum leap for out control and MIS needs. It works and is here now. No excuses. We have sufficient computer power, memory, software and bandwidth beyond any immediate need. Time to get to work...

Part III — Software:
Indispensable Exasperation

A long transition for manufacturing managers to become computer literate is nearly complete. Generations growing up with computers are now well into the manager ranks. Manufacturing software packages have improved so that the term "user friendly" is now nearly accurate. Applications using math techniques within real-time control constraints are fulfilling the promise of better products.

Battles over PC operating systems for industrial applications continue. Windows NT and CE have available software advantages; Unix and versions of it are still alive and are well proven. VXI used with the fast VMIbus system is also still performing well. Now the Unix offshoot, Linux, is stirring much interest with its open availability and ease of Internet interfacing. Stay tuned!

Articles Comprising Part III:

"Significant progress doesn't come from the formal planning process of an American corporation, it comes from a couple of guys doing something that hasn't been set down on a list"

—*William G. McGowan,*
CEO, MCI

Building a Solid Foundation

An architecture is any ordered arrangement of the parts of a system. Just as with a building, so with computers and other microprocessor- based systems; without due attention to the fundamentals of construction, a computer system will not withstand the rigors of the environment in which it is placed.

A system architecture is a structured set of protocols (connections) that implement its functions. Functions are what computers, communications network and manufacturing systems do. They are the properties of a system's components. In a building, components might include mechanical and electrical equipment, acoustics, illumination and traffic layout, in addition to the structure itself. For software, the components are the fundamental functions, and their interconnection is the architecture.

Reuben S. Jones of Softech Inc., Waltham, Mass., and I are involved in several efforts to establish standards for the architecture and operating systems of modern real-time systems used in industrial automation and process control. Among the many groups concerned with implementation of real-time systems, their operating systems and message protocols, are NCMS (National Center for Manufacturing Science), DARPA (Defense Advanced Research Projects Agency), ISA (Instrumentation Society of America), ICS (Industrial Computing Society) and IEEE (Institute of Electrical and Electronic Engineers).

If you've been a regular reader of this column, you'll recall that the three determining characteristics of a real-time system are that it be robust, predictable and fast. That is, fast enough to do the job. For bank check clearing, that means overnight. Credit cards must be verified in seconds. Scientific experiments need nanosecond response. In the control and automation realm, responses in milliseconds are sufficient.

Robustness determines the availability of the system. Reliability may be considered one aspect of robustness. Often ignored, but equally important, is predictability of performance. The system should consistently respond in the same way and in the same amount of time for a given event. This response must be independent of all other events and actions requiring the attention of the system.

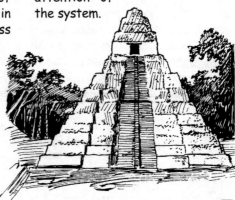

A layered architecture is one of the simplest and one of the most robust possible.

A "simple" architecture is one that can be understood by the user community. Tremendous efforts have been expended on formulating automation standards that ended up being so complex that they were never implemented. Exquisitely elegant engineering solutions won't fly if the business case can't be made. We can't ignore the needs of users and operators.

One possible architecture of a microprocessor-based system consists of layers that may communicate and service each other only in a vertical mode. A layered architecture is one of the simplest and most robust architectures possible. But not only layered architectures exist. Some other architectures are the autonomous and the distributed, as well as the four basic computer designs. Computer scientists also talk about "crystal" and "mud" concepts.

The "layered" approach can be applied to systems both large and small and to both hardware and software. An upper layer may use only those services provided by the lower level adjacent to it. Lower layers provide service only to the next higher level. Complete stacks may be wholly replicated throughout the system, provided layering rules are maintained. Partial replication is only allowed if a common base is used. Vertical communica-tion is allowed, but partition-to-partition communication is not allowed.

An analogy to this system of rules can be found in the hard-wiring of electrical and electronic systems. We know we should not arbitrarily install "jumpers" that defeat portions of an electrical control system. Likewise, a micro-processor-based system shouldn't have jumpers between the innards of a chip and the workstation dis-play. Although obvious when stated in hardware terms, many of us, myself included, do the equivalent of "jumpering" when designing sys-tems and software.

As a result, we get in trouble when specifying communication protocols and computer operating systems. Large scale, single-user implementations such as distrib-uted computer systems and com-puter-integrated manufacturing systems are particularly suscepti-ble to suffering negative impact from ad hoc approaches to design.

Design discipline and a sensitivi-ty to the need for a sound founda-tion should be foremost in the implementer's mind when embark-ing on any project. Build on a solid foundation and Murphy might stay home.

Brochures, Bibles and Baseball

Over the years, I've been involved with nine software-based companies. Some of them did not succeed in the marketplace, but the technical stuff came in within budget. By that I mean no overruns of more than 40 percent. One advantage I had was that I ain't no software guy. My background is physics. Tough to know everything about nothing. Ah, but I digress.

In any case, I think my experiences with software can be applied to a lot of different businesses. To my mind, the secret is to start at the top and cycle down. The actual software program coding is the last thing done. The first task is to write a sales brochure. This establishes the basic offering of the product. The brochure is what gets buy-in from the engineers. Everyone must sign off on the project goals and product definition. The most difficult sell is always the internal one. If you don't make that sell, you'll encounter the enemy. The enemy is anarchy. The brochure answer the age-old question, "What businesses are we in?"

Then comes the "bible," a 200-page tome in which you describe the details of the implementation, including types of work stations, operating systems and platforms. The author is usually a high-IQ landscape artist with some moss on the north side. The bible is written from a marketing viewpoint, not a sales or technical outlook. It is, however, written for the implementers and quality people. It's not done by committee, but lots of interaction is needed to get it done. And it can take more time to do than the actual coding. Although not a perfect document, it must have no fatal flaws.

Everyone must read and sign off on this internal document. It must be generic enough to meet all of the needs of the user, without constraining the demands for new features on the part of sales. From that point on, all questions can be met with the same answer: "Is it in the book?" No extras or changes are allowed once the project is underway, unless the change fixes a fatal flaw.

I know of a company that always has three projects underway and never finishes any. Their product vision is so short term that the chosen technology is invariably obsolete in 18 months. And anarchy rules there, with the engineers blackmailing the management in order to achieve their own nefarious ends. Thus, a "new" project is launched every six to nine months, incorporating the latest chips, etc.

Managing an organization is a

The brochure answers that age-old question, "What business are we in?"

classic engineering problem. It is not an art form. Marketing in-put, brochure, bible and implementation—in that order. Feedback is frequent and substantive. Organization is flat and loosely coupled. Teams of two to five people are assigned components of the bible.

Management is achieved by watching resources, not objectives. In other words, staffing levels are a given, and the project prices must conform to them, not vice versa. Meanwhile, the moss-covered grand old man circulates, all the while giving forth little gems of wisdom. If any of the star techies rebel, immediately discard them. Baseball teams built around stars seldom win the pennant.

Let's take an example. Assume the assignment is to do a networked financial transaction package. The first offering is a desktop blister pack. The Bible is written and calls for use of three-man teams. The segments include user entry, database, optimizer, communications, reports, screen and management. Each is treated as a product for sale. Each has testing and quality procedures. Each segment is defined well enough to be subcontracted to outside suppliers. About 30 percent of them will be. The secret to software management is to manage the process, not the project. The first pass

should be a prototype and treated as a throwaway. As for the engineers, give them all the tools and tell them when they are done.

As each piece is completed, it must be signed off on by the moss-covered guru, marketing people and focus groups. The metrics of implementation must be rigorously adhered to. Early field testing is done by the internal users and selected forgiving customers.

As I've said before: software is someone else's idea of what you want to do. Test and maintenance is 80 percent of the task. The problem is, small companies have too many chiefs and not enough Indians, while large companies have too many Indians. Discipline, small teams and attitude will make it happen.

Better Than the Real Thing

As you may know, I'm a member of the CIO (Computer Integrated Operations) committee of the National Center for Manufacturing Sciences (NCMS), Ann Arbor, Mich. Recently we had a CIO meeting at Kodak to decide the fate of nations. And I pontificated hugely about modeling and agents. Jim Heaton and Van Dyke Parunak beat me soundly about the head and shoulders because, as usual, I was locked onto the latest Morley romance, and would listen to little else.

Associates say I have only about 20 percent actual content in my talks. The problem is, they don't know which 20 percent it is. So they have to listen to the whole lecture. Even over the dinner table or on the ski lift.

Anyway, my talk centered around the several ways modeling can be done.

One way is to make assumptions about how simple agents interact with the environment. The product of those assumptions is then examined for comparison with the real world; in other words, an "experimentally" developed model. Controls engineers observe the actual process and tweak a model until the simulation behaves the same as the real process. The model may be only the starting point for model construction.

Often, changes introduced by engineers are motivated more by intuition, experience and hunches than by academic "soundness." The proof of the pudding is the behavioral correspondence to the real systems, not the approach.

Examples of experiment-based model include the "Bullet" train in Japan. General Motors has a truck paint shop that is optimized for pull-through scheduling and process control using this approach. So-called generic algorithms embody the experimental approach since they make random changes to a model, observe their effect and integrate successful changes into the model. Artificial intelligence and artificial life studies also use experiments in real time, as does life itself.

An older mechanism for modeling is the deductive method. This approach is common in research settings. It depends upon the accuracy and thoroughness of our theoretical understanding of the physics of the process. Theories are, at best, only a rough approximation, and a theory-based model may work only for simple situations. A model based solely on the-

"Associates say I have only about 20 percent actual content in my talks. The problem is, I don't know which 20 percent it is."

ory may not, in fact, represent what actually happens in the process.

The third approach is the usual compromise. Inductively we can monitor the actual system with new math techniques. By reconstructing the geometry of the system, engineers have a model based on behavior, rather than theory. This approach has been applied extensively to formally chaotic systems, where theory is invariably inadequate to predict performance.

In general, model-based control incorporates a computational model of the process into the control system, rather than just reflecting the parameters and architecture of control. This approach appears to be more robust over a wide range of parameters. It does require that the engineer have an explicit model available.

Over the last several years, NCMS has funded Delphi studies for ascertaining the near future of technological thrusts for industry. Software is at the head of the class. All indications for the future involve modeling and/or agents. These studies herald an ongoing revolution in software. As with the hardware in the computer revolution, we can expect to see modeling in control systems change from an art form to a true engineering discipline supported by the pillars of scientific understanding.

Thanks to James Heaton and Dr. H. Van Dyke Parunak for their input in putting together this column.

Ladder Logic, Sci-Fi Wise

I need a life. Too many of my idle hours are spent on the 'net. But my time is not totally wasted. Some browsing in the automation bulletin boards indicates that ladder logic, the PLC language, needs some work. The original ladder logic was discovered (invented?) in the late '60s by the group that started Modicon. We sought to introduce real-time, user-friendly software to industry. The only available languages at that time were based on icons or flow charts.

One priority in designing the interface was that it had to be intuitive, for use without training. The only significant change since that time has been the addition of cells, or function blocks.

Defined one way, ladder logic is a symbolic representation of what amounts to a control circuit. The power lines form the sides of a ladder like structure, with the program elements arranged to form the rungs.

Expressed another way, ladder logic incorporates computational agents with convergent transforms. The database is composed of outputs from agents and sensor inputs. The agents examine the blackboard database independently for subsequent execution. It is predictable, inherently robust and fast enough.

Rather than abandon ladder logic in favor of computer pro-gramming languages, such as Basic or C++, we should make the first significant upgrade to ladder logic in several decades. By incorporating a range of suggestions, we can resist the temptation to yield to the beckonings of "Windows 99."

To do so, a committee of august citizens will be drawn from the MAP, Pentium, Denver baggage, OMAC and Waco teams. The committee will be charged with incorporating into ladder logic such things as virtual reality, holographic databases, floating point math, public key encryption for I/O and full preemptive multitasking.

The new ladder logic will allow extensive subsets. Nothing will be defined to allow each user the greatest possible freedom. It will be the first politically correct computing language, with blame always assigned early in each

To encourage others to join in helping make the reality a dream.

implementation and enforced by a standards committee. One potential, unforeseen benefit of this approach is that each company will finally have a proprietary standard to ensure user loyalty.

Technically, ladder logic is NOT a "go-to" language. It is more "come-from" and, in the Boolean sense, uses the "if-then" concepts. But these are modern times and we can today do better. Suggestions gathered by means of an informal survey include incorporation of such concepts as:

- What if — to allow decisions before the events take place;
- True, if false — a contradiction mode;
- Run for each — apply this output to the entire plant;
- Or else — also referred to as the Mafia instruction;
- Who else — the modern I/O call;
- Why not — an ideological run-time routine;
- Run fail-safe — OK-to-fail mode;
- Limbo exit — where your programs go when unwatched;
- Don't do — execute and remove all traces of programs;
- Won't do — will do only if the instruction "May I" is typed; and
- To be or not to be — self-destruct mode.

Many such improvements are being considered by the committee. In addition, we will look at traditional uses of ladder logic. Function blocks are typically timers and counters. This can be taken to a whole 'nother level. For example, resumes can be automatically generated based on the density of bugs in a given program. Or, overhead presentations can be generated based on project overrun statistics.

Finally, since some seem to think factories will one day be run by word processors, we'll counter-attack "Word" dominance with "Ladder," thus ensuring an unlimited future for the MBAs of Cambridge. We're also petitioning the U.S. Department of Defense to send the language, free of charge, to our friends in Iran and Iraq.

Job creation resulting from support of these standards will be greater than any Porkware project ever imagined. Much like any other standard creating body, we will measure progress by number of spec sheet pages produced each month. At the conclusion of years of work we will be in position to thank the committee for the work done thus far, and to encourage others to join in helping make the reality a dream.

Going to the Show

The nightmare begins on Prytannia Street in New Orleans, where I'd gone to attend the Instrumentation, Systems and Automation Society (ISA) convention, the biggest U.S. show on process instrumentation and control. The problem was that instead of being booked into a businessman's hotel, I'd somehow ended up in a bed and breakfast! Therefore, no data port, no CNN and no turbocharged air-conditioning; I had to retire without my usual e-mail fix.

Sleep finally comes, but slumber is disturbed. I dream I'm walking the second floor—the aisles extended endlessly. As I walk the walk, each and every person pitches their product: "Better, cheaper, faster." Each and every product consists of a large sign and a color monitor, and numerous attendees stand stunned and awestruck before the arrays of icons displayed on thousands of screens. Soon, the body types, apparel and even hair pieces of the booth attendants begin to seem identical. I walk faster and faster. There seems no way out—and I need a way out! But suddenly, I'm awake again and trying to find my way in the dark… Drinking too much coffee late at night will get you every time.

My nightmare did not, in fact, come true. ISA does an excellent job of managing one of the largest industrial conferences. We experienced no problems, despite the threat of a hurricane.

At the show I noticed a lot of:

- virtual-reality head sets;
- interest in the future of technology;
- acceptance of the digital world;
- serious attendees; and
- good Cajun food.

One of my appointed tasks at the show was to present the Modicon Rising Star Award on behalf of the Industrial Computing Society. The recipient of this award must be less than 35 years old. In an era in which so many of those with engineering backgrounds go for MBAs and move to marketing or management, it's nice to give recognition to those who stick to

I had to retire without my usual e-mail fix.

the technology side. Make the move to marketing and you get only money, but stick with engineering and you get plaques.

This year's award went to William Staib, designer of the Intelligent Arc Furnace (IAF) Regulator, a neural-network-based electrode positioning system for electric arc furnaces. The production IAF system, released in September 1992, uses the pattern recognition abilities of neural networks to predict and correct for changes in furnace operation. Continuously adopting its control strategy, the IAF is said to save several million dollars a year on a typical installation by increasing productivity by over twelve percent and by reducing power consumption and electrode wear.

To control an arc furnace is to control a 10 to 60 megawatt lightning storm. Because the IAF is among the first proven, large-scale industrial neural network applications, the system has been heralded by both the steel industry and by the neural-network research community. Staib, who graduated with an M.S. in electrical engineering from Stanford University in June 1993, is presently developing neural-network control applications for other industries. We can only hope that Staib continues to contribute to technology development.

Staib did get a $1,000 check in addition to his plaque.

As one ages, one becomes the recipient of more and more awards. The work one does over a lifetime adds up. If you've chosen your parents well and manage to outlive your peers, you will eventually be regarded as a "smart guy." What no one knows is that you just didn't have enough gumption to ever change careers. But guys like Staid are different. His award is based on the speedometer and not the odometer. More young achievers should be recognized, and more attention needs to be focused on the speed of the journey and less upon the miles traveled.

An Interview With Myself

What do you mean by the "demise of software"?

Software will not go away. The phrase "demise of software" simply means that the bloom is off the rose. Software mystery and magic will soon disappear from the terrain of intellectual investigation. The same thing happened not that many years ago when the electric motor labs were dismantled. In the 1800s, motors were imbued with sense of wonder, and vice-presidents concerned with standards, central single motors or distributed power were the rage.

Electric motors have not gone away. And neither have candles or horses. But our intellectual fascination with them as unexplainable objects is not as close to the surface as it once was. We can expect that in 20 years the care and feeding of software will be such that it won't need to be extensively taught in the universities.

Talk about manufacturing at the point of consumption.

Historically, carbon units (humans) have located manufacturing efforts at the point of raw material—smelters at ore deposits, refineries at oil fields and pottery kilns near sources of clay. Large manufacturing facilities have been located based on labor availability considerations. Location at the source, not at the point of need, has been the paradigm.

Recently, in the paleontological sense, our society has become consumption constrained, as opposed to production constrained. As a result, how we think about production and consumption will change. We must locate innovative replication facilities at the point of consumption, to make quantities of one unit with zero defects. Zero inventory is the result. Enabling technologies are desktop manufacturing and the Internet.

In fact, teleoperations will allow companies to ship a virtual product—i.e., the information needed to manufacture—rather than the real one.

Taking this idea to its limits, strawberries will be grown in the supermarket, automobiles assembled at the dealer and perfect-fit jeans sewn at the clothier. This may seem farfetched at this point, but the trend is clear. Desktop publishing, Turbotax and L.L. Bean are all examples of changes being wrought in the wresting process.

The manufacturer of the next millennium will either have distributed physical facilities at points of consumption

Strawberries will be grown in the supermarket, automobiles assembled at the dealer.

or make extensive use of specialized express services. Assembly performance at the airport or trucking warehouse will become commonplace. Product design will in large part be governed by replication and distribution considerations. Even pharmaceuticals will be tailored to the individual. The automated drugstore will arrive hard on the heels of the World Wide Web.

The term bandwidth seems to mean different things to different people.

Once many humans learned to read and write, the oral tradition of the great memorizers was replaced by the medium of pen and paper, offering greater flexibility in design. The publishing industry perfected distribution via high-speed physical conveyances, so great was the need for information.

Now, electronically, it takes but seconds to send a page from one side of the nation to the other. The rate at which you can transmit the page, however, is determined by bandwidth. The number of letters per minute, the number of bits per hour or the numbers of cycles per second, are all mechanisms of bandwidth.

We can decode, calculate, word process and work in color over long distances. Computers now have, for all intents and purposes, infinite capacity. The wire connecting the two computers, however, is narrow and constrictive. It cannot transmit all the code in one computer to another computer in less than an hour or two.

We can expect in the near future that there will be a one-hundred times increase in the channel width among computers worldwide. This will eliminate all the electronic hesitation and puts the lag in communications back where it belongs...in our own heads.

Still greater changes are in store for software, manufacturing and communications over the next several decades. The winners will be those who embrace what's staring us in the face. Good luck.

Much Data, Little Knowledge

This columnist was recently a panel member at a medical conference on data mining and knowledge discovery. I know little about the subject, but agreed to speak anyway. Fools rush in...? It was actually "an invitational conference for collaborative development of a network-based, real-time rule-server (NRR) for timely decision support."

Physicians need tools to help them make timely decisions while in the presence of patients. The complexity of medical science is such that the unaided mind cannot cope with the challenges of delivering state-of-the-art care. Doctors need to mine the knowledge base quickly enough for the knowledge to make a difference. As I listened to the introductory speaker make these simple points I was struck by the similarity of what he said to explanations I have heard vis-á-vis the need for manufacturing execution systems. Only in this case the process is a patient.

Now in the medical field most of what's currently found in databases concerns financials— billing and costs are what's important and what's been put on line first (much as the first uses of writing were for shipping manifests, not literature). Useful tidbits related to treatments, such as patient history, pharmacy records, surgery schedules, admitting forms and lab test results await integration into the computing realm.

It won't be long before 50 gigabyte databases are fairly common. It's easy to be buried by that quantity of data. Search engines, WWW published sites, terabyte memories and T1 communications bandwidth on every desktop can be problems as well as solutions. But, as with every evolutionary challenge, new tools are being developed in response to the tensions arising from immense available storage capabilities and the promise of the Internet, which is the largest database of all.

The task is to reduce the data to a small quantity of useful information. Knowledge data decision (KDD) automates data analysis by identifying valid, novel, useful and understandable patterns in the data being examined, using data heredity and parentage tracking techniques that have themselves been available for two decades. The technology base for KDD is also available, if expensive. Rapid database and Internet access, including WAN and LAN TCP/IP compatibili-

Finding information where no one has ever gone before.

ty, is mandatory. Enabling technologies include neural nets, parallel systems, emergent and inductive software, landscape theory and system understanding.

Many of the applications envisioned for KDD presuppose the ability to access a variety of data formats and procedures. This implies either a single standard for all databases or using a technology for heterogeneous access. Look for the latter to be the reality. Java and applets are but a start in this direction.

In fact, there is other important data that will not be found in any database. The knowledge engine also needs awareness of the present state of the "process." For example, one of the "secrets" of the PLC is that the system state is treated as part of the database. Real-time capabilities (fast, robust and repeatable) must be part of the decision-supporting process. Obviously, it is better to make a less-than-optimal decision at the right time, as opposed to a potentially better decision too late. This applies to a surgeon in an operating room or a planner trying to formulate the next day's production schedule.

Imagine a system that operates by connecting diverse databases to a variety of search engines. Interfacing is done via memory-mapped blackboard systems across several embedded PCs with a rules-based scalable, management operating system coupled to it. Applications in manufacturing might include:

1) understanding complex purchasing behaviors;
2) demand management and forecasting; and
3) sales-channel analysis.

Additional applications can be envisioned in transportation, automotive, semiconductor, pharmaceutical and food industries.

KDD is more than a search engine. It is a knowledge-based repository that helps in decision making by identifying trends, behaviors and patterns too evanescent for less compute-intensive engines (ourselves) to handle. With the Web being the *de facto* standard for databases of the future, we'll be able to put the execution system, and even the entire enterprise, on a phone jack. The opportunity is there, we have the data; now we need the answers. All we lack is the courage.

The Emperor's New Clothes

Manufacturing execution systems are a kind of middleware for manufacturing. A term originally coined by Advanced Manufacturing Research (AMR), Boston, an MES is the means to bridge the gap between process control and production management systems. In addition to being the means to better integration, the MES includes specific functionality for scheduling, quality control, document management and others.

The history of MES began with applications developed by specialists aimed at tackling specific problems. Tracking operations, in-process materials management and scheduling are some of these areas. Coupling this functionality with a relational database management system, for both real-time and archival applications, is crucial to effective utilization of MES.

Users, quoted in the trade press, cite productivity improvements of up to 50 percent for some operations. But it's no secret that while some enterprises are getting real improvements, others have struggled to realize benefits using the "traditional" MES approach.

As middleware, an MES must interface with shop-floor process control applications as well as with enterprise resources planning systems (ERP). As most readers of this magazine are well aware, ERP is a transactional system that typically includes functionality for financials and accounting, human resources, manufacturing and distribution. ERP systems are increasingly involved in supply-chain management.

My experience has been with the factory floor proper. But we need to understand the management aspects of the floor, factory and enterprise.

As middleware, an MES must interface well with the floor as well as the upper mainframe functions. Historically, companies patched together execution systems without even knowing what they were. Then came the promise of a single, integrated system. But costs, and the inclusiveness of the functionality provided with the system, raised questions.

We expect the future will bring a matrix style of applications suites to the MES concept. In this way, an overall framework allows each vendor to work within the matrix, offering fully compatible components to an open system, but still allowing

Some people are more mercenary than others.

for innovation in implementation.

Why is the title of this column "The Emperor's New Clothes"? It all started with an aluminum-tube journey from New York City to Manchester, New Hampshire. My seatmate was a high-paid consultant with a blood alcohol content above the legal limit. He lauded the existing structure of many information systems, since it assured consultants of an obscene profit for a long, long time. Need for customization, difficulty of installation and excessive procedural code were all included in his praises. Installations that took up to a year seemed a glorious thing to him.

Listening to my seatmate brought to mind the story of *The Emperor's New Clothes*. In that story, the emperor was conned into believing that the finest clothes could only be seen if the viewer had a pure heart. Otherwise, the fabric was invisible. The con men took the money and skipped. When the emperor traveled in a parade wearing nothing, a child asked, "Why is the emperor wearing no clothing?" Since the heart of a child is pure (except in L.A.), it's obvious that all could see that the emperor was truly traveling in the buff.

In the angel investment business, we always ask about exit plans. The future is hard to foresee, and we are interested in the ability of the company to change business vectors. Plant software should have the Darwinian aspect of survival and quick adaptation. Software that can only feed off of eucalyptus leaves is vulnerable to cusp events. Flexibility is not a key feature of complex, consultant-ridden systems.

Vendors are smart and working hard to respond to these criticisms. The new silicon and the ADSL/WWW communications will certainly enable systems in a way unimaginable today. Web and Internet coupling are, so far, only now being recognized in the manufacturing applications space. The oscillating fashions of distributed/centralized architecture continues. Work at the Santa Fe Institute suggests that the technical answer is a hybrid of both the distributed and centralized approaches. We shall see. Remember Unix?

So watch out for interoperability, web compatibility, costs of install and change, and buzz-word decisions. Fast and flexible response to the business environment requires a behavior business model, not a procedural model. The future framework concept and hardworking vendors are our ultimate solution. If the business fits the model then go for it. Otherwise, keep your heart pure.

Beans for Breakfast

eorge Markowsky stopped by the other week. He's a computer science professor at the University of Maine, and an old friend and co-speaker from the Chaos and Cybersea events. Over breakfast, I was forced to confess to him that my late morning phone interview with Sun Microsystems would be a sham.

"What will the subject of the conversation be?" asked my would-be rescuer.

"Java and PLCs," I replied. For anyone who doesn't know, PLCs are to factory automation what PCs are to office automation. What's more, your columnist is ofttimes credited as being one of the inventors of the PLC. But my background in physics makes me one of the world's worst programming experts.

Since I was the PLC guy and Markowsky was the Java expert it was clear that I would be paying for the eggs and pancakes.

What the *&*&#$$% is Java anyway?

Java is more than a programming language. It is a software platform due to the Java virtual machine, which simulates a computer in software. The virtual machine can run on PCs, other computers already on-line or on hardware specifically for Java.

According to George Gilder, the technology guru, "Java [is] an efficient pro-

gramming language that is safe, simple, reliable and real-time, yet familiar to anyone who has used C or C++...it is interpreted rather than compiled—that means it is translated line-by-line in real time in the user's computer rather than converted to machine language in batch mode by the software vendor... [Java] is compiled not to an instruction set peculiar to a particular microprocessor, but to a virtual machine or generic computer. Putting the language into an intermediate binary form allows creation of programs that are not locked into any particular hardware platform, but can still be adapted to run fast."

I asked, "Can't that be done with 'any' language?"

"Yup," Markowsky answered.

Gilder says the difference is, "Java can be 'compiled' line-by-line in real time. That is, it is interpreted in byte-level code in the client machine. It means programs no longer have to reside in the machine where they are used, or be written for that machine in order to be executed by it."

Of course, one man's idea of "real time" may be quite

ILLUSTRATION BY PAT LETOURNEAU

55

It became clear to me that what's needed is a real-time virtual extension to Java.

different from another's. What Gilder is talking about is real time in the database sense, not for purposes of process control. For factories, real time has to be "fast enough, robust and predictable." A real-time machine cannot have time-dependent multitasking.

It became clear to me that what's needed is a real-time virtual extension to Java. It would have to be small and compact with all the manufacturing process dependencies built in. Markowsky and I agreed this may not be possible with the present set of tools. The industry may need a silicon Java chip set for real-time applications.

Since then I have checked with others that concur there are hurdles to real-time applications of Java, including need for real-time Java virtual machine and a load time compiler. They point out that there are many good, inexpensive development environments for Java code, that cross development is the name of the game and that we can leverage trends in network development for own goals.

There seems to be general agreement that as a development environment Java still has a way to go, but that it holds great promise. What's revolutionary is what Tom R. Halfbill pointed to in a recent article in *Byte* magazine—Java's higher level of software abstrac-

tion. Java code is free of microprocessor and operating system constraints. This freedom is the future of computing.

Before breakfast was over Markowsky and I decided our real-time extensions—both hardware and software—to the Java virtual machine should be named EXPRESSO. It may be that others have already usurped this moniker. We don't care. We started off the project in the usual way. Before I sauntered back to the Barn for my phone interview, we ordered some coffee cups with the name emblazoned on the side.

We're on our way.

Part IV —
Managing Manufacturing:
Multitasking Personified

I mprovements in manufacturing came from the mass production operations throughout the 1900s. Such manufacturing methods are the fundamentals for today's manufacturing engineers. In the past two decades,
manufacturing managers have had to add to their professional routines.

First, they need strong computer skills. They should understand computer capabilities in I/O technology, CAD/CAM, software programs, tele-
communication interfacing and networks. In addition, manufacturers must
stay with advances in materials, processes methods and new product intro-
ductions. They also need to relate directly with suppliers and customers.

"He who has never made

a mistake is one who

never does anything"

—*Theodore Roosevelt*

Quality From the Start

The manufacturing process itself is only part of the quality story. Once upon a time, quality in U.S. manufacturing was a controversial issue. Lack of it allowed Japan to get a leg up on us. But quality in manufacturing is a given now. It's an expected feature of any manufactured product. If you don't have it, no one is going to buy what you make. Cars and TV sets are expected to work.

Quality has got to start right from the design concept, and has to extend until the technologically obsolete product is deposited in the junkyard of history. That's the issue today. It's a total life-cycle consideration. And as a designer, that search for quality in the total life-cycle has got to be my passion.

Three years ago at Flavors Technology, we finished a large-scale design for a real-time super-computer. Five hundred megabytes of memory and more than a hundred large-scale microprocessors were installed. The boards were almost two feet long and contained many layers of circuitry.

The design also incorporated object-oriented hardware concepts. Circuits could be energized and tested in small parts and later consolidated into the whole machine. This allowed the design to proceed a segment at a time and the computer could be manufactured in the same way. The whole is a complex assembly of simple design and test elements. Conservative values in loading and timing are used throughout. Diagnostic testing was built into the hardware and software from the start.

Results? Significantly less cost in design and manufacturing. We like to think that the parts could be put into a basket and shaken. The components would soon settle into position and the run light comes on. Even the system architect (me) could bring up the system.

Another example of quality in design comes from automotive arena. Cadillac redesigned the Seville's rear bumper. The number of parts and the assembly time were cut in half. The benefits of this reduction exceed any accountant's estimate. Less time and fewer parts mean the statistically analyzed quality of the manufactured bumper is significantly higher than that of the previous design. Any repair service is simple and replacement parts are cheaper. It's the customer who realizes the benefits of life-cycle quality. And

Do SPC, SQC, quality circles and concurrent engineering really help the customer if no one in the service department will answer the phone?

Cadillac benefitted from a quicker design cycle and lower costs.

I'll tell you where else our quality control efforts have to extend to—our own tools and factories. Factories should be built on the basis of 50-year cycles. This means that we have to be increasingly agile. But how else can we successfully deal with technology and process innovations? Tools, instruments and process concepts have to be freed from antiquated facilities. In the future, factories must be designed to allow for gradual but continual improvements to the manufacturing process.

Let's take a trip into the future. Factories will be making small lots to order. Flexible tooling is what will make this possible. For example, an engine facility might produce a proliferation of types with from four to eight cylinders. A large variety of displacements and configurations will be made with the same agile tooling. The same facility will also meet aftermarket demands with the remanufacture of motors and major components. Routing of both new and recycled products through the same facility, reuse of tooling and equipment—that's what will reduce manufacturing costs and ease pressures on the environment as well. Makes you think, doesn't it?

My point is that we too often consider only the quality inherent in the manufacturing process. It's the buzz word *du jour*. Besides design, which precedes manufacturing, we must also look for quality in customer service. Do SPC, SQC, quality cycles, QMC, concurrent engineering and ATE really help if no one in the service department will answer the phone?

Recently, my wallet was stolen. Inside were all the credit cards in the world. Immediately the card people responded, canceling and replacing the cards. They were responsive and quick. That's quality. To hold the customer like a king is part of quality efforts. I will cancel almost any normal business appointment for a customer issue.

The U.S. had a quality problem and we were building crap. Now we're building "good crap." But it's time to move on and become the leaders in total life-cycle quality. Design, distribution, service—all performed with agility—are integral parts of the quality process. We've established manufacturing quality. Now let's have quality that starts with the design philosophy and never stops.

Winning With an Agile Defense

Agile manufacturing is the ability to quickly and at little cost reconfigure a production environment. The concept's advocates recognize that the prospect of unanticipated change is the only constant. To embrace change, you gotta change your system concepts.

The Department of Defense (DoD) has already decided to concern itself with agile manufacturing. You might consider this an unexpected bonus of the peace divided. DoD's having less money to work with is changing its procurement technology. The revised strategy will emphasize new product science and technology as opposed to large-scale production. The plan calls for continued feasibility demonstrations to advance new technologies, while limiting manufacture of production quantities, unless deployment is needed.

But while there are distinct advantages to emphasizing research and deployment, it's inevitable that the production base will shrink as a result. Industrial plants will be re-tooled for non-defense industries. Production expertise in weapons systems might prove to be unavailable when the

need for it arises. To avoid a big-time scramble mode when defense production capabilities are again needed, the DoD knows it has to be able to rapidly respond. Agile manufacturing will provide the DoD with the rapid response capability it needs.

In the private sector, agility is a priority, just as it is for defense provision. For Detroit to win its economic battles, it has no less need for agile response capability than does the DoD. Incorporation of technologies that enable an agile response on the part of manufacturers and their suppliers will assure success in both defense and in international commerce.

What will these technologies enable? They will allow technologists to rapidly integrate concurrent-engineering efforts and to change systems and products across the entire supplier community; to ramp up quickly in the face of a sudden need; to rebuild quickly in the face of downsized facilities and lack of trained personnel; and to engage in small- and single-lot production without significant cost penalty.

The needed technologies and their associated manage-

ILLUSTRATION BY PAT LETOURNEAU

In some respects, industry is ahead of the research community on this important issue.

ment practices are already here. What technologies are they? Well, historically, machinery was customized and software was generic. Now it's the reverse. The "iron" is generic and the software is highly application-specific. Software defines the system. The new technologies of Object and Symbolic manipulation are here and available.

Significant progress is being made in advancing agile manufacturing concepts. In addition to my own work in parallel processors at Flavors Technology, Mark Roth of Allen-Bradley is working on autonomous module architectures. Roger Schappell from Martin Marietta is working on schedulers. David Greenstein from GM is already demonstrating a very flexible manufacturing system. Steve Benson reports that Digital Equipment has defined generic encapsulated manufacturing modules for electronic assembly. My own mentor for agile systems has been Rick Dove of Paradigm Shift. A lot of the info in this column was supplied by him. Benson's work with industry, the National Center for Manufacturing Science (NCMS) and the Defense Advanced Research Projects Agency (DARPA) will, we anticipate, lead to the deployment of truly agile systems that have application from the factory floor to the total supplier infrastructure.

Very briefly, there are a number of issues that adhere to the topic of agile systems. First, the need for agile systems must be recognized. There exists a certain amount of mistrust between academia and industry, and in some respects industry is ahead of the research community on this important issue. Unfortunately, the track record for those brave few who are the first to deploy a new technology has been very poor at times. Finally, the perceived costs of implementing agile systems have to be looked at very carefully.

Agility in manufacturing systems will reduce start-up times and costs. Agility will enable a robust supplier network. Surge capacity, an important market issue, will be significantly improved. Once effective models for rapid development and deployment are fully articulated, the DoD will be able to design products without going into full production. Should the time come, God forbid, the DoD will, however, be able to ramp up quickly.

Many Challenges Face Systems Integrators

What are systems integrators? Are they glorified resellers? Are they employees of the manufacturer? Or, to put the best face on it, are they "technology partners"? In the purest sense, a systems integrator delivers turnkey solutions for projects outsourced by the client. The integrator provides expertise in hardware, software and communication tools needed to complete the contract per the user's specifications.

Manufacturing is a unique arena for the adventure of a systems integration project to play in. Unlike banking or insurance, each manufacturer is a distinct, whole supplier to the market. Therefore, each client is different. The integrator must also know something of the client's industry, and the technologies used in the plant's processes and functional areas. Familiarity with the client's business culture and customs is always a big help.

The systems integrator also assumes financial responsibility for the job and, as part and parcel of that, responsibility for the technical performance of the installed systems. Like a building contractor, a systems integrator takes full responsibility for all the subcontractors. In short, the integrator serves as the overall project manager, the single point of contact, assuming all the praise and all the blame.

Markets for systems integration services continue to grow. Projections for revenues in 1993 are in the $80 million range, indicating growth of about 15 to 18 percent. While current economic conditions have probably caused the cancellation or postponement of some large projects of over $5 million, enthusiasm for medium-size projects continues unabated. This enthusiasm tends to be strongest in companies with annual revenues of $50 million to $500 million. Larger companies are holding back and using in-house services wherever possible.

It is in the arena of contracts of $5 million or less that the open systems and reengineering initiatives are being played out. The open systems and reengineering concepts are there to remind us that technology is not the solution.

ILLUSTRATION BY PAT LETOURNEAU

The modern systems integrator needs to define a niche market wherein his expertise can be clearly demonstrated.

Technology, and computers, are the tools.

Just having computers is less important than how they are used and how their systems are maintained. That's why it's so disappointing to see how persistently users continue to make buying decisions based on price alone. I see some life-cycle costing, but lamentably little. Outsourcing, though, is continuing to grow in popularity as a strategy for reduced labor costs in almost all industrial sectors. Today, real men outsource.

Two-thirds of the growth in systems integration markets will be attributable to the proliferation of software for manufacturing and the growth of markets for real-time computing. What's driving growth in this sector of the market is the complexity of both the tools and the processes involved. The intricacies of networking and open systems also contribute heavily to the need for specialists.

The trend toward specialization and division of labor will continue. The modern systems integrator needs to define a niche market wherein his expertise can be clearly demonstrated. With more and more of the hardware vendors, including the very largest computer companies, taking up the business of systems integration, it is incumbent on the small, niche players to emphasize their strongest suit—process expertise.

I wish systems integrators would take a page from the advertising business. In fact, many of them already have. The relationship between the client and the server is paramount. It is not defined by the duration of any one project. For the systems integrator, the term "client" suggests a special relationship. The role of the integrators is to propose specific vehicles for meeting clearly defined strategic goals. Perhaps in the future more integrators will be on retainers to their clients.

If you are a systems integrator, define the market that you serve. Look there for your market-share growth. Define a bumper sticker for your services. Make sure you're centered on the process, not the technology. In other words, emulate reality. The year 1993 will be a good one for the systems integrator.

Dream On...Again

When flying at 40,000 feet, terrain becomes as much texture as anything. Only the huge features are discernible...the Grand Canyon...New York City...irrigation circles. Individual aspects are subsumed as a pattern emerges. This emergent pattern may not be obvious to the sidewalk plodder.

The view from up here is that software is fast becoming the most important part of any manufacturing process. Brand new factories are equipped with software to the tune of between 10 to 20 percent of the total costs. The single largest component of the Defense Department's procurement budget is systems software and its subsequent maintenance. Even Detroit's urban transit "people-mover" has more software/silicon than steel content.

But still we dream on. Meetings in autoland still center on mechanical aspects of the product even though combined revenues of the companies listed in the *Manufacturing Systems* Top 50 are larger than those of the entire programmable controller (PLC) industry.

Looking at the list got me wondering about the realist vs. the dreamer and what effect small,

innovative companies have on a general market. A dreamer is someone unconnected to reality. By this definition, Einstein, Edison and Ford are realists, while some of today's managers and politicians dream on, hoping we'll eventually regain the past.

The realist is the person who accepts the world for what it is: continuous change.

I know of a computer model that demonstrates how big change can be introduced into market by means of small differences. Using Visual Basic, a grid of 64 companies was constructed. A random number represents the company size. Turbulence is introduced into the system by changing the quantitative designator of the smallest company, replacing it with another random number. Each of its four neighbors are changed in the same manner.

Eventually, after many cycles, even the largest (strongest) company is affected. Entropy demonstrated within the system proves interesting. There are at first long cycles of stationary (non-movement) changes. Then sudden jumps at other locations in the grid occur. These appear to be cyclical.

Now comes the

ILLUSTRATION BY PAT LETOURNEAU

Individual aspects are subsumed as a pattern emerges.

fun. Even when we temper the changes and populate the grid with large companies, the dynamic characteristics remain the same. Those dynamic characteristics seem somehow immune to minor algorithmic changes in the paradigm. We call this the "Guano Syndrome." What will happen will happen.

A. Strelzoff of Modicon/AEG did the programming. Gotta give credit.

This simulation model seems to parrot the real world. Many of the companies in the Top 50 weren't around even a decade ago. But these small companies with their innovative products are changing the industrial landscape. And change will occur independent of efforts to delay it. To stay competitive we must embrace the viruses of the future. They are the agents of change.

Many of the ideas incorporated in the model came from studies done in biotech and artificial life. Artificial life simulations model the Darwinian process and lead to an understanding of how relationships impact a changing environment. Relationships between predator and prey leave little room for behaviors that respect the individual. Likewise, change in software markets is often unrelated to what the individual companies that make up that market originally intended.

So those often labeled dreamers are actually the realists, and vice versa. Realists seize upon the moment and ignite the fires of revolution. Such a revolution is seen in the Top 50 list, and in plenty of other places, too.

Get a Life

I was one of the best dressed at the Artificial Life IV workshop held at MIT in July. I try to have a science learning experience at least once a year. Even if I don't learn something, I learn something.

But because it was an academic gathering—with only a sprinkling of press and industry types—there were very few suits. As I glanced around me, I felt real pride in knowing that the heels of my socks were on the bottom and that I knew what to do with my tongue when I wasn't talking.

The weather was hot, but it was good to be back at the old alma mater for the first time since my 40th reunion in the spring. Alumni visiting campuses have always been the bane of the students: We tend to grin idiotically.

Anyway, if you are wondering what artificial life is, the workshop proceedings say it's a term that covers "a range of computational ideas concerned with attempts to synthesize phenomena normally associated with natural living organisms. The media for these synthesize phenomena normally associated with natural living organisms. The media for these synthesis experiments include computers, robots and (bio)chemical soups."

Life is the continuous adjustment of internal systems to the external environment. Artificial life involves self-evolving objects based on silicon platforms.

Humans are carbon-based. Since artificial life would be an extension of ourselves, it might be misleading to label it "artificial."

Whatever is alive is bent on survival. But the rule isn't so much survival of the fittest as rejection of the nonfit. The mechanisms of change include mutation, crossover (sex) and natural selection. At the workshop, these principles were demonstrated by means of graphic simulations, video presentations and robot demos.

Speakers stressed the difference between species learning and individual learning. Artificial life systems can be made to express either, or both. It also seems that "evolutionary" change occurs in spurts, rather than along a continuum. Much occurs in 10 generations, and then nothing for a thousand generations.

Another characteristic noted was that groups of simple life forms can coalesce to form other,

I knew what to do with my tongue when I wasn't talking.

more complex forms. In some sense, humans are a bunch of spare parts flying in close formation. Every type of cell has the same basic "programming." Or, to take another example, cells that make up termites are alive, as are the termites and as is the hive formed by the termites.

Why am I interested in artificial life? After all, I'm a manufacturing and computer nerd. But if we can reduce software complexity and improve manufacturing processes via artificial life, then I'm all for it. Industry-related topics discussed at the conference included:

• behavior in self-organizing production systems;
• self-learning assembly systems;
• market simulations and economics;
• survival characteristics of large companies;
• robotic control; and
• autonomous vehicular systems.

Although actual applications are so far limited, I foresee widespread use of artificial intelligence technology in the next decade or two. Think about where the use of robotics was 10 years ago. At one time it was considered "far out." Today, we take factory robotics for granted.

Enabling technologies needed to fulfill the technology's promise include a "Darwin chip," i.e., one capable of "evolving." "Swarming" architectures for dealing with large systemic problems are being worked on by the Santa Fe Institute, Flavors Technology and the Japanese, among others. The next ALIFE conference will be in 1996 in or near Kyoto, Japan.

Some of the bullet items I carried home:

• learning is useless in a stable environment;
• stop directing change and let it happen;
• activity (not function) is the important thing;
• problems and solutions are the same;
• formality is wrong;
• organisms need competition; and
• simple rules are sufficient.

But the best lesson of all—

• learn or die...

The Eyes of the Beholder

A reader writes: Ask five blind men to describe an elephant, and you will likely get five different answers that, when combined, may not look very much like an elephant at all.

Which reminds me of the *Saturday Night Live* skit—back when it was actually funny—of Chevy Chase and Jane Curtin arguing whether or not a new aerosol product was a floor wax or a dessert topping. As the huckster selling the product, Dan Akroyd was quick to point out that it was both—floor wax *and* a dessert topping.

What does this have to do with manufacturing? Well, think of how people view reengineering; the most recent addition to the toolbox that includes JIT, TQM and other techniques we use to invigorate our companies and the way we do business. Is it an operational approach, or is it technology? Does reengineering mean better information technology, or fewer levels of hierarchy in the organization?

Several years ago, manufacturing at Modicon, the PLC-maker located in North Andover, Mass., was productive, but certainly not as efficient or lean as possible. There were production lines and storage strategies dedicated to specific product families. Capital equipment was duplicated across these lines in a manner that led to waste, and the production facility wasn't delivering the customer response sought by upper management. Some manufacturing was being done offshore to take advantage of lower labor costs, but the offsetting problems in raw materials, work in progress and finished goods in the inventory pipeline was staggering. Add to this the challenge of quickly and effectively executing an engineering change order, and it became clear that something had to give.

Modicon recognized the problems and took action. Manufacturing and service were brought back to the United States. All manufacturing for all product families were drawn into a new state-of-the-art facility. The facility was built for Modicon with attractive lease terms. The company introduced new storage, material handling and warehousing methods and technologies to streamline manufacturing, applying lessons from Kanban systems and pull-based scheduling.

ILLUSTRATION BY PAT LETOURNEAU

Ask five blind men to describe an elephant, and you will likely get five different manufacturing facilities.

They introduced work teams, giving responsibility for a product family to a single group that cross-trained. Suppliers were drawn into the plans as well. The list of vendors was trimmed by more than 70 percent, and those who remained were asked to participate in the quality effort. Modicon enforced participation by tying performance to payment, and rewarded it with loyalty to the vendors. All incoming quality control was effectively eliminated.

What does all of this have to do with elephants, dessert toppings or reengineering? We've taken many groups from the U.S., Korea, Japan, Europe and elsewhere on tours through that manufacturing plant. What you learn over the course of many tours is that depending upon who leads the tour, you get a different view of the elephant.

Financial types will focus on the cash angle: better use of capital equipment and a new facility with a long-term lease helps the company better serve the customer. The manufacturing guys talk about product flow through the plant and worker empowerment. Operations people will tell you about vendor programs, storage strategies and increased turns. The sales guys tell you it all works because the plant uses Modicon products that you, too, for a price, can have as your own.

In reality, reengineering is all of these. Sure, they are using state-of-the-art technology in surface-mount work cells, using vision, expert systems and other technologies to their fullest. Yes, the transport system that moves product from point A to point B is controlled by modern automation products. Yes, anytime you can get a good deal on your lease, it's great.

But taking the longer view, this is reengineering.

If you asked the folks at Modicon if they've reengineered, I'm not sure what answer you would get. My guess is that whatever you got would have a trunk, but from there I won't make any conjecture.

—Bob DeSimone
Flavors Technology

When Products Ruled the Earth

President Truman paraphrased an old truism when he once stated, "The only thing new is the history you haven't learned." Within the history of technology development, a cycle of initial complexity, subsequent simplification and eventual reappearance of complexity has often been seen. This pattern is evident in the history of the automotive industry. The semiconductor industry also is at a particular point in its cycle, with complexity set to make a comeback.

The thesis is that for any given technology revolution, its beginning is announced by the appearance of product-oriented companies serving narrow niche markets. Consolidation follows, mostly due to market and financial pressures, which results in large conglomerates offering generic products. Eventually, though, these conglomerates do market extremely diverse product lines. They do so by use of sophisticated process, rather than product, technologies.

When the automotive industry first appeared, the diversity of companies and products was bewildering. Each "instant" entrepreneur made cars for a niche market. Then, Sloan and Co. purchased many of these "silos" of excellence and consolidated their efforts into what is known today as General Motors.

Note the name. "General" products—single offerings—that attempted to satisfy the entire market. Simple automobiles designed to establish a market, rather than respond to it. At the time, market manipulation was the fashion; focus groups and market response studies were not done. The "Big Three" told the buyer what was needed, not the other way around. As Ford stated, you could have any color you wanted, as long as it was black.

Today, no two cars are identical, at least potentially, because modern automation and information technology allow wide-ranging product variations. The manufacturer is closer than ever to responding to market demand in real time. Technology lends a greater voice to the user.

This is true in other industries, too. Time turns many business concerns offering custom products into consolidated companies selling mass-production models. The same consolidated companies eventually offer many custom prod-

Semiconductor manufacturers must take a page from the discrete manufacturing handbook.

ucts. The circle is complete: Today's factories are market-driven, not product-specific.

Semiconductor manufacturing is poised to enter this last, market-driven phase. Today's semiconductor facilities are product-specific and cost a billion dollars each. Think about it. A billion dollars is spent equipping a facility to meet the forecasted *future* demand for a chip it will produce.

But Moore's Law states that chip density increases by a factor of 10 every five years. We are now at 0.2 micron technology with 10-million-junctions-per-chip. In the early 1970s, a megabyte of memory cost a half-million dollars. Now the cost is about forty bucks. How long can this go on?

Single-product semiconductor factory efficiencies are decreasing as the market develops, thereby delivering less bang for the buck. Semiconductor manufacturers must take a page from the discrete-manufacturing handbook: make agile factories that resist obsolescence. To take the example of the auto industry again, the trend is away from transfer lines to more agile CNC-machining centers automated with robotics handling.

The end result of agile manufacturing is that we can become managers of the present and not the future. Managing the future always causes headaches. Agility and flexibility are the watchwords of any commodity-like technology business, not any one specific product. The semiconductor factory of the future will be centered around the process of manufacturing, not the specifics of the product.

The transition will offer many opportunities to process-equipment vendors. Rather than the academic-oriented equipment found in existing semiconductor facilities, Silicon Valley will engage in agile execution similar to that found in other maturing industries. This will reduce the cost of new products. Wall Street valuations placed on semiconductor vendors will drop to reflect realistic expectations of revenue and margin, not hype. Silicon Valley will learn from the Rustbelt.

Leap of Faith

We all have to deal with reluctant management, especially when trying to sell a new idea. When selling that idea, understand that your goal and the organization's goal may be two different things.

First, don't be too original. When presenting your new baby to critics in the company, be aware that incremental change is the watchword of the successful internal champion.

These incremental steps might include a "toy" concept demonstration, full simulation, interactive hands-on and finally, the alpha units. Alpha unit designs are meant for the garbage heap, not eventual full-scale production. For example, in the early days of the PLC, we did a modular design which did not include programming with ladder logic. Then the Morley Rat Pack did a whole new unit called the 084. A languishing performer, but it did attract the attention of some MBAs, who threw me out of the company. Eventually, the 184 concept made all of us a couple of bucks.

The secret was to sell "relays in a box" and not a new computer with a real-time operating system. The point is, work within the given infrastructure. Don't try to build a new world. I actually remember the weekend that we forced all the Rats to eliminate the "C" (computer) word.

You can distort the infrastructure, but not destroy it. Try promoting evolution as a means to a more radical mutation. Hide behind the cloak of merely trying to "improve" an existing state of affairs.

The early days of Andover Control were a maelstrom of the latest technology in a sea of flagging sales. We convinced an engineering meeting to ascertain the reasons for flaccid market response.

We first listed three reasons for purchasing our products: havoc, cost and performance. But after much anguish, we ended up with comfort, status and safety as the key features. The technical staff then used these criteria as the philosophical underpinnings of any design.

We were off and running.

Too often, we decide based on rumor and not direct experience. Is RISC really faster? Does the Pentium really have a problem that affects us? Is Chicago the answer to everything? Prestige, PR and media exposure tend to make silk purses out of sows' ears. But we can use this to our own

The market for potatoes is marginal.

advantage. We can teach them what to believe. Too many engineers scorn the hype of the salesperson, and then fail to understand why their anti-gravity proposal is rejected.

Recently, we invented true anti-gravity and demonstrated it to some venture capitalists. The demo was a floating aluminum plate. The plate hovered over the conference table, supporting a coffee cup.

The guy said, "So what?"

Agitated, I ranted on about changing the world we live in and other small benefits. He finally recanted his initial rejection and asked me to give an example of a concrete benefit.

"Making long-haul trucks able to carry more cargo," replied your intrepid columnist.

"What cargo?" he countered.

"Potatoes!" I frothed.

"The market for potatoes this year is marginal, we'll never make money that way."

My lawyer says that the guy will recover and my court costs will be minimal. Best of all, the story illustrates my point that striking out to change the world will only mess up your average presentation.

Making the idea serviceable in the minds of the presenters is the challenge. If acceptance is desired, approval should not entail risking the company or the careers of the backers.

To sell your latest idea for a double-diphthong modifier widget, consider the following:

- talk pain, not benefits,
- beauty is a real issue,
- it should make market sense,
- sizzle and PR sell and
- frame and power sell.

Most new inventions invent themselves. Inventors are merely lucky bystanders. Mutations occur in nonlinear jumps and fall into place. Revisionist history makes the issue "obvious," but most of us know better. The inventor must be irrational, and not listen to conventional wisdom. But most of all, the time must be right.

Sense and Prejudice

I recently viewed a television show about so-called "junk science." That's the kind of science that happens when what's nothing more than opinion is dressed up in the credentials of a recognized expert, thereby becoming the received wisdom. The resulting nonsense assertions have several distinguishing characteristics. They mysteriously appear, spread rapidly, tell a good story and are mostly false.

In this way, science begins to be a "world explanation," as magical as any ancient mythology. After all, no one of us is able to confirm that the world actually is made up of subatomic particles, and precious few will ever get how quantum mechanics works.

Yet the success of scientists in predicting and controlling atomic events means many have faith in what science asserts. Unfortunately, that has led to blind acceptance of some pretty sloppy thinking. Take Social Darwinism, for example, or the belief that a cure for cancer is just around the corner.

We, the techies, are a debunking bunch. Or so we think. As engineers and managers, knowledge workers and software developers, we—who have faith in progress and the rational—have our own fetishes, just like any communal tribe. We have our own "lite" version of junk science: a kind of "junk wis-

dom." You can evoke the parameters of our fables by interjecting any of the following into a conversation: cellular phones, diesel exhaust, the transputer, Lisa, 68060, NASA O-rings, iron in spinach, computers and productivity, travel is glamorous, nobody uses Macs, a paperless office, sexless engineers, Ramen soup and Jolt, Unix wins, Unix loses, paradigm shift or linearity.

Too often, engineers and managers take actions based on the assumption of competence, and are surprised when they don't get results. Since we are the experts, we avoid scrutiny. We state to anyone within Java range that the data speaks for itself, and that all data is wisdom. Ergo, since computers are data manipulation devices, computers can solve all our problems.

Let's level with each other. We know that data doesn't speak. Real-

Do we really ever make contact with the "objective" world?

ly. Honest. What we see always is based on our preconceptions. The old saw, "seeing is believing," has it backwards. It really should be, "believing is seeing." Ultimately, that's as it should be. At the beginning of this century, Ernst Mach and the Vienna Circle tried to come up with a science based solely on empirical facts. They soon found that without theories to provide a provisional ordering of facts, no knowledge is possible. What's frustrating is that once we admit this, we're forced to wonder if we really ever make contact with the "objective"" world.

Bringing this discussion back down to earth, *Dilbert* is probably the supreme unmasker of management's unstated assumptions. But we'll try our best to identify a few.

Fables of manufacturing include:

• lowering costs is the key to prosperity;
• mass production lowers costs;
• lowering overhead decreases costs;
• individual behavior can be modified by the corporate culture; and
• management science exists.

Marketing fables include:

• product features and product benefits are the same thing;

• packaging is the product;
• ignorance outranks intelligence;
• lower prices sell product; and
• if you don't know what you need, you don't need it.

These memes have a longevity and vitality beyond all understanding. Perhaps the human spirit seeks to expand its mastery, even at the expense of truth. We must rationalize and improve products and performance *ad infinitum*. A treadmill indeed. Our training says we're logical, and that the Newtonian clockwork is the be all and end all of our existence.

Few of us question the day-to-day morass of legacy prejudices: The bosses are idiots. Smart people are weird. Work is dehumanizing. The old days were better. There is no end to what technology can achieve. We are at the top of the food chain.

Now, don't get me wrong. I often believe some of the above. But I try at least to occasionally question the *a priori* assumptions of my own existence. Can't we all just be a little more honest with ourselves?

Room for Improvement

This is a story about the real costs of running a manufacturing enterprise. Imagine, if you will, a medium-sized manufacturing facility somewhere in the great state of California. This hypothetical facility replicates designs (i.e., it manufactures) for the industrial controls and systems markets.

As all modern enterprises do, this facility strives for profit. To aid these efforts, it tracks costs. Its accounting methods attribute costs to "legacy"-type departments such as marketing, production or engineering. Just as there are silos of information, there are silos of cost. This unfortunate practice leads to selfish localized optimization to the detriment of the whole organism. Reducing local costs makes heroes. But each time manufacturing "reduces" costs, engineering takes on more of the burden.

Factory-floor labor accounts for less than 10 percent of costs in a modern electronics factory. Looking at the entire building, administration and marketing costs run on a par with that of the plant floor. Engineering costs, however, frolic at Dilbert-ish levels of more than four times floor costs.

It seems that manufacturing costs have been brought under stringent control, except those for technology support, including maintenance of the facility and product design. Both evolutionary and novel development costs are skyrocketing. The budget expands each time a new buzzword enters the front gate.

Technology management issues are fast becoming the only bottleneck in any agility strategy. Several years ago, I spent the day at an ultra-modern automobile facility. With me were three "skunk works" people. Two airframe specialists, an automotive guy and a computer guy (me). At the end of the day, our observations were as follows:

- Skunk works are two times faster than large organizations.
- Skunk works are five times less costly than larger organizations.
- Skunk works build single-target items and not broad band products.

ILLUSTRATION BY PAT LETOURNEAU

Does that mean that the right reorganization could reduce a $100 million dollar budget to $20 million? Very unlikely. But there's lots of room for improvement. Much like inventory turns going from four-to-one to six-to-one. Outsourcing, agile manufacturing and enterprise management have brought significant focus to plant operations. One more to go: engineering.

There are some good books around on these subjects. Three of the best are *Mythical Man Month* by Frederick P. Brooks, Jr.; *Fixing Broken Windows* by George Kelling; and *Organizing Genius* by Warren Bernis and Pat Ward Bedermen. Behaviors advocated in these books include the following:

• hire top people;
• have strong leaders;
• deliver a message from on high;
• have an enemy;
• ship product;
• view work as the reward;
• advocate complete local control;
• form small work groups of no more than seven;
• avoid surprises;
• use metrics of performance;
• keep outsiders out; and
• base pay on performance, not longevity.

My list is a lot shorter:

• be the dumbest person in the place;
• have the best tools;
• stay focused;
• don't pay very much; and
• tell them when they are done.

How do you know when you have a good shop? What are the metrics? In 1994, we did a rump study for NCMS on metrics for technology groups. The indicators are:

• money spent on new projects;
• cited patents;
• percent of revenue invested in technology compared to others in the same segment;
• people turnover; and
• does anybody want the output?

The title "Fixing Broken Windows" alludes to an interesting psychological phenomenon. If a warehouse has a broken window in a seedy part of town, all the windows will soon be broken. If we fix the first window, others are much less likely to be vandalized. Disorder begets disorder. And in engineering, low esteem encourages low productivity. That's why powerful computers, free Mountain Dew, keys to the building and less meetings are all so important.

Technology and management concepts must be devised to increase agility and productivity in the implementation and support function we call engineering.

Rule the World

alk about complexity and chaos as scientific principles may not receive as much attention in the popular media as they once did several years ago, but the ideas haven't gone away. My staff and I continue to host workshops on the application of chaos to manufacturing problems. We run two a year, one in April in Santa Fe and the other in October in Boston. Overseas conferences also are hosted—these are better attended than our own American conferences.

Dr. John Holland was the keynoter in Santa Fe last April. He is professor of computer science and engineering at the University of Michigan and is a pioneer in the field of complex adaptive systems, and inventor of the genetic algorithm. An excellent speaker.

Normally, I attend the entire session. This year, however, I had to leave the conference to give a complexity talk in Orlando to IBM. The big boys are more and more interested in these new sciences and philosophies.

I left immediately following my Chaos 5.0 tutorial on the first day. I drove to Albuquerque from Santa Fe, stepped into the airport and flew to Orlando via St.

Louis. I gave the talk at the Orlando Hotel and returned to Albuquerque—all this without ever stepping outside. All the travel, land and air, was done in aluminum tubes. The hotel was at the terminal. This fascinating process—travel from Santa Fe to Orlando—took just 24 hours. My attendees hardly knew I was gone.

I personally like the small entrepreneurial companies, but the big boys have been showing up at the conferences as well. Attendees at this conference included representatives from the U.S. Air Force, GM, Ford, Babcock & Wilcox, CSC, Corning, Deere, Eli Lilly, E&Y, GE, Honeywell, Intel, IBM, Los Alamos, MIT, Monsanto, Oak Ridge, Foxboro, Wonderware and Yaskawa. Why you ask?

The technical community realizes that over the past decade, production systems have grown more complex. And with this complexity comes an exponential growth in the complexity of control mechanisms. Top-down, classic architectures have run their course. Attention has turned to the bottom-up approach of agent-driven emergent systems. The thirst for real mechanisms to solve complex problems is at hand.

ILLUSTRATION BY PAT LETOURNEAU

The thirst for real mechanisms to solve complex problems is at hand.

Many popular articles have described some of the new approaches. Professional societies have sessions on complexity in some form. We are not in total agreement as to terminology when it comes to the new control means. You hear terms like autonomous agents, responsible agents, emergent systems, chaos theory, genetic algorithm, fitness landscapes and strange attracters all bandied about.

The Ernst & Young Center for Business Innovation hosts an annual summer conference—this year in Boston on August 3 and 4. Speakers included Stephen Jay Gould, Stuart Kauffman, Chris Meyer, Mitchel Resnick, John Casti, Tom Petzinger and your humble columnist. Wow. Such company. The workshop emphasizes shared conversation in addition to presentations.

E&Y has also formed a company with Dr. Stuart Kauffman, Bios, to commercialize the science of complexity and complex adaptive systems for industry. Bios is modeling large systems that include multiple plants and markets in dispersed locations. The goal is to understand and model resource allocation, yielding low-cost adaptation to changing landscapes. As Bob MacDonald, Bios CEO, states, "This is a tall order, but complexity science offers the potential for new insights and recommendations."

At October's second annual Boston workshop, the keynoter will be George Kelling, author of *Fixing Broken Windows*. He will discuss the recent success in crime prevention accomplished by means of small catalytic changes in individual behavior. This may help us to understand how small changes can impact our own organizations, as was discussed in last month's column concerning increased engineering productivity by means of catalytic management.

Other practical subjects to be covered at the conference include scheduling systems, sales-force agents, process control, the World Wide Web, IPR and chaos.

The idea that controllable systems can be better understood is taking hold. We, in manufacturing systems, can improve all aspects of our enterprises through understanding the process that the real world imposes upon us. The idea that fractals, flocks and factories are related can bring us significant productivity improvements. Learn or die.

Part V — Enhancing Corporate Wealth

Sales of manufactured products create wealth. Improvements in business procedures decrease costs and enhance corporate wealth. It should be recognized that these basic wealth-generating activities are technologically- and people-based.

Rapid advances in computers and telecommunications have squeezed corporate decision times to uncomfortable levels. The successful enterprises have adjusted with better IT capabilities and by flattening the hierarchy. The Silicon Valley dotcom operations are the wave of the future—regardless of recent stumblings.

"What we are trying to
do is put the soul of
a small company into
a big company's body"

—Jack Welch

Technology Development and Market Creation

In the U.S., technology transfer has traditionally proceeded from academia and the military research environment to industry. That's changing. At a recent Defense Advanced Research Projects Agency (DARPA) workshop, industrial sector attendees were asked about technology transfer. They replied, "Sure, we're interested in technology transfer. What d'ya wanna know?"

There's plenty of talk going around about technology development, technology transfer and revitalizing commercial innovation. Too much of it is just talk. But the issue of corporate support of technology development is real enough. We need investment in technology because it creates wealth in a way that investment in things like increased market share never does.

While technology development and transfer has been done well by some government and quasi-governmental agencies, the creation of markets is almost exclusively the province of private venture capital in pursuit of new technology, and the creation of jobs is almost exclusively the province of companies with under 500 employees.

A recent story in *The New York Times* illustrates some of the pitfalls of technology development in the U.S. Kingsbury Corp., a Keene, New Hampshire-based machine tool company, wanted to use a new type of diamond-coated tool it had obtained for testing from the National Center for Manufacturing Sciences (NCMS), Ann Arbor, Michigan. But no suppliers produce the new cutter in commercial quantities.

Why couldn't Kingsbury buy a better mousetrap? According to the article, niche technologies often go begging for development dollars, and even those attracting investment tend to mature slowly. Why didn't tool companies rush to invest in making the new cutting tools? Reasons cited include questions about performance, cost and where it will fit into the supplier's product strategies. Suppliers may fear better technology approaches are due to appear. Attaining full-scale production may be technologically or economically challenging. Finally, patent uncertainties and questions about the potential market inhibit investment.

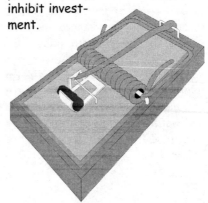

Why couldn't Kingsbury buy a better mousetrap?

Too much business focus in the U.S. bears on the short term and too many management decisions are made by the technologically illiterate. Technologists need management champions if their projects are to see the light of day.

To make it happen, the champion needs the following: a business plan; evidence of customer interface; good people; a significant barrier to entry by competition; 100 percent of a zero market; and an intriguing and fun-filled adventure. The last may seem frivolous. It's not. Projects centered around boredom and drudgery seldom succeed.

To industry's credit, there've been some countervailing indicators. For example, there's been a recent trend for automotive companies to promote engineers in the senior management ranks. And their strategies for 2001 focus on systems: the automobile as product and manufacturing as process, together as a single system.

The statistics, however, can be discouraging. For example, I am associated with the Breakfast Club, a group of investment angels in southern New Hampshire. We figure that only about one in 25 technology initiatives ever becomes tremendously successful. An additional five might prove worth the money invested, but the rest are worthless. Venture capitalists find a 20 percent success rate astounding.

In my experience, a realistic target for full payback of an investment is seven years. For the first five years, it's seldom apparent whether there's a real chance for success. Staying power and adequate funding during this period is very important. Unless you give good people enough time and money to do the job, you're backing failure.

What kind of organization best suits technology development? Development teams or "skunk works," like those Lockheed used developing the SR-71, are a corporate approach to technology development that works. Skunk works take approximately half the time do a project, and for one-fifth the cost. There is a caveat however. Generally, a skunk work yields souped-up race cars, not Chevrolets.

Industry must use skunk works, give technologists clout by promoting top management from engineering backgrounds and budget research adequately. Technology generates markets and creates wealth. Everything else is just reallocation or redistribution of wealth. We must understand what that means and look to technologies 25 years in the future, and to what their market will be.

Chasing Chaos in Santa Fe

just got back from Santa Fe, where I discussed with a number of like-minded individuals—in an informal conference setting—chaos and complexity theories, especially as they relate to issues in manufacturing. Nearby was the Santa Fe Institute, a multidisciplinary research and graduate education enterprise involved in studies of the principles that determine the dynamic behavior of complex systems.

Models of complex systems seem to have application in many fields, including biology, computer science, economics, physics and political science—hence the multidisciplinary approach. Besides institute members, attendees included representatives of Rockwell International and Allen-Bradley, a Rockwell company, and others.

The kind of ideas we talked about go by many names—agent-based systems, emergent programming, chaos theory, nonlinear dynamic systems and artificial life. Basic to many of them is the notion that total systematic behavior can arise from the union of localized systems.

For example, the classical way of thinking about something like an ant colony is to imagine that somewhere there is a master plan that governs it. But, in reality, no one "programs" the ant colony as a totality. What really happens is that the relationships of each ant to all other objects—to include its environment and the other ants—taken together define the colony. Individual ants have rules they follow, and the interactions of the rule-governed individuals end up determining colony behavior.

The same applies to human societies. Rather than imagine there is one essential "invisible hand" that guides the market system, it may be more productive to envision any number of "hands" that interact in a somewhat characteristic fashion to form an economic system of such complexity that is in many ways unknowable. In fact, the prejudice that there is a single, right answer to a question of group behavior can be very dangerous.

What's true for macroeconomics systems is also true for manufacturing systems. To apply these theories to automated production, what we do is replace scheduling techniques that treat the manufacturing system as a totality—in

Basic is the notion that total systemic behavior can arise from the union of localized systems

which all variables must be accounted for—with techniques for parallel scheduling of random and chaotic processes. In Santa Fe, we at Flavors Technology presented our emergent system approach to scheduling paint booths at a General Motors auto plant.

In this application, the booths are the defined objects and the programming code describes the booth's relationship to its environment. A "schedule" per se is not programmed, but only the objects and the rules of the interaction. The result: a flow of trucks emerging from the autonomous agent called the paint booth. Each time we run the simulation, the results are different.

Use of these type scheduling techniques should be considered whenever a system is complex, i.e., large, complicated and interactive and intractable. Most complex systems are too...well, too complex to be wholly conceptualized. This alternative approach focuses on operational variables and is much better able to deal with unforeseen eventualities. Practitioners continue to be surprised by the results they get. Systems seem to get better in startling ways. Considerable savings in software development are also possible.

As with anything new, acceptance in the engineering community has been tepid at best. Apparently though, Japan is an exception. Recently, Japan hired some of the top researchers from the Santa Fe Institute to study chaos theories in Japan.

At the conference, institute representatives gave the attendees a solid overview of some of the new techniques. We also visited the institute during afternoon tea and wound up giggling a lot. Presentations by Dove, Patenak, Roth and others updated us on work currently being done in applications in neurons, global system behavior, adaptive learning and social interaction.

In general, the attendees were enthusiastic and interested. In fact, we elected to work late one evening rather than frequent the bar. We also decided to do the whole thing again next year and to start a news letter. An e-mail net is being set up by Ken Crater of Control Technology Corp. on internet [ken@cthulhu.control.com]. He has information available.

Achieving Total Simulation

We're at the point now, I think, where most engineers work with computer-aided design/manufacturing (CAD/CAM) systems on a daily basis. Originally meant to speed design and reduce reliance on paper, CAD can now do 3D visualization of a fully rendered design. The dynamic characteristics of a product or process are also handily dealt with on the most modern platforms. What CAD is to products, simulation is to processes.

The technical envelope for simulation is being pushed by factors found outside the realm of manufacturing systems. Once breakthroughs are made, though, it doesn't take long before the latest developments are applied to the manufacture of goods.

The military recently demonstrated before congressional committee a virtual, real-time, nationwide, battlefield simulation. Many types of simulations contributed to the results and small segments of "reality" were included as well. The program's purpose is to train personnel and to test tactical response and new weapons concepts. Reality segments provide the "unexpected response necessary to validate the process."

Computer games sold to consumers generate so much cash that the industry is willing to invest megabucks in product development. Consequently, computer games have become tremendously sophisticated in a relatively short period of time. Simulation of evolution, the management of a city or the behavior of an ant hill trains players to interact with representations of reality, i.e., reality as process.

How can this notion of games as a model of reality be applied to manufacturing? Let's design a "SimMan" game that allows the players to simulate manufacturing processes such that interactions are possible in faster-than-real-time. Bringing the game to bear on the design, testing and manufacturability of products will enable lust-to-dust analysis of the product life cycle.

Before we begin, what are some of the problems with simulations to date? One big problem is validation. Not one manufacturing scheduler has so far gained wide acceptance in industry. This is simply because random events and sporadic occur-

How can a game, as a model of reality, be applied to manufacturing?

rences spoil the best-laid plans of both mice and men. The simulation process needs to be certifiable. Speed of interaction should never be an issue. The simulator should be faster than what is modeled. Sometimes I think one of the big advantages the Japanese enjoy is that they don't apply MRP II widely.

What are the benefits of playing such a game? "What if" questions can be answered quickly. With a city simulator, the mayor can be mayor for over 200 years within a span of several months. (Who says that time travel isn't possible?) For manufacturing, costs in product and life cycles can be ascertained. Automotive companies can weigh the expense of quick model changeovers against the marketplace advantages to be gained thereby. Agile manufacturing and small-lot concepts can be reality-checked.

Today, we continue modeling with hard and fast assets. In other words, too often we build it and then we think it through. First the factories are built and the products are made, then come the surprises. By then, changes are difficult if not impossible. Further down the road, responses to changes in market forces become expensive and time-consuming.

Therefore, the simulation/game must prove valid over a wide range of possibilities. It has to be played in simulated time, at least an order of magnitude faster than the process itself. It should be extensible into marketing and aftermarket situations. The interactive model should examine hard and soft costs charged to the manufacturing process.

Using military or game simulation techniques, we aim at lust-to-dust scaling. Validity is attained by inserting pockets of reality throughout the model. The pockets occupy about one percent of the model and might include segments of the manufacturing and supplier structure.

Connecting real running machines to the simulator increases the validity of the modeling process. Via distribution across a wide geographic and management landscape, scheduling gets a reality check. Scalability allows players to test small segments of the manufacturing process or expand into design and marketing at a later date.

Of course, the final step will be from simulation to reality. It's been recently suggested that the line between simulation and control is becoming blurred. In the near future, simulation will be control. Not only will what-if questions be answered, but simulation will also serve to clarify how we can "make it so."

Reengineering...Art or Science?

They tell me that in 1993 the theme for *Manufacturing Systems* is reenginnering, acheived mostly by means of better use of information technology. During the year, many business publishers issued book titles, and innumerable consultants gave seminars to teach managers and executives the theoretical lingo linked to this, by now, slightly soiled buzzword. But don't get me wrong. Just because there's a buzzword involved doesn't necessarily mean there's no substance behind it.

The old ways (pre-reengineering) were based on economic assumptions that no longer hold true. Cutthroat competition in high-tech and all other industries has left profit margins razor thin. The ability to move assembly work offshore means that to justify manufacturing in the U.S., productivity must be maximized.

Want some examples of the old ways? Grow the empire, dump the problem, keep your head down, keep the boss happy and "today is the same as yesterday." Under the old rules, customers never repaired their own equipment. Local warehouses were the means to access the marketplace.

Manufacturing planning was done to satisfy stocking needs and was not the means to monitor and service the order stream. In short, a legacy of paper and lack of empowerment.

But it's the revolution in communications, fostered by the computer-as-commodity, that makes new productivity gains possible. It's the explosion in software applications that makes it possible to question the old way of doing things.

The computer industry acts as a catalyst, making reenginnering possible. At the same time, it has lately been a prime target for reengineering efforts, due to the competitive nature of the business. Stamford, Conn.-based Gartner Group, a market research firm, recently released its market and financial analysis of the largest 100 worldwide revenue producers in the U.S. computer hardware, software and service industries.

According to the report, the good news is the rebound in revenue growth to seven percent for U.S. technology vendors from 1991 to 1992. The bad news stems from drastic reductions in gross profit margins for computer hardware, especially large systems. The "Yardstick" forecasts rev-

Reengineering should allow us to serve the customer, create value, own the problem, listen to the user and constantly learn.

enue growth in 1993 at "only slightly better overall," despite continuing cost-cutting measures, especially in selling, general and administrative costs, and in manufacturing budgets.

PC vendors fared better in 1992 due to their ability to reduce costs faster. The Gartner report also indicates migration to client/server architecture is accelerating, with gains attributed to PC price elasticity and performance improvements. But while PC supplier revenue grew a little more than 13 percent, their products were more than five percent less profitable than the year before—at almost 33 percent of revenue—due to price wars and cost cutting.

Stop for a minute and check out the trend lines highlighted by the report. Continued growing demand for information technology, but at the same time, technology breakthroughs and increasing competition make turning a buck tougher than ever. Growing popularity of client/server, the computing paradigm that best suits reengineering. And some beneficial effects of reengineering seen in the reduced price of PCs, caused as much by streamlining of marketing, manufacturing and distribution, as by technology gains.

Reengineering should allow us to serve the customer, create value, own the problem, think team, listen to the user and constantly learn. The rule is "pick it up only once." The benefits to be had through reengineering are actually much greater than you might reasonably expect. From my own experience in reengineering processes used in software development, I can say that they are of an order of magnitude.

The rush to reengineering is not without its problems. Change is always messy, takes longer than anticipated and may have unintended consequences. It's very important to remember, too, that once something has been downsized—all too frequently a key component in reengineering efforts—you've lost capabilities that are very difficult to regain, should the need arise.

Chaos 3.0

Many emerging issues in complexity and chaos have actually been around a long time, to put it mildly.

All is flux, nothing stays still.
—Heraclitus

From the ancient Greeks to modern architecture, patterns of complexity have influenced the arts and the sciences. And they'll continue to do so, in the philosophy of science, which seeks to explore the relationships between theories and empirical data; the social sciences, still looking for appropriate systems of measurement; and even manufacturing. The National Center for Manufacturing Sciences (NCMS) already has several projects that explore the impact of emergent systems on mass production.

More conventional notions of how systems work have always assumed such things as linearity between inputs and outputs, but whether in biology, chemistry, physics or factory control, there are instances of sharp, wild-card disruptions of linear patterns. Yet we remain convinced:

In all chaos there is a cosmos, in all disorder a secret order.
—Carl Jung

Recursive iteration is one of the most important tech-

niques being examined as a means of solving theoretical problems. Linearity and non-linearity are both being examined with this iterative process. The use of probability theory, and its incorporation in computer algorithms, has led to applications that take into account systems that produce behaviors unexplainable if linearity is assumed. Modeling of real-time systems is but one application.

Yaskawa Corp. has a full-running model of the "bullet train" for the Japanese railway. Each segment of track, the vehicles, station performance and power distribution are all part of the model, which operates in nonlinear fashion based on emergent analysis.

In an article recently published in *Industrial Controls Intelligence*, Bob DeSimone of Flavors Technology points to why the existence of unforeseen stimuli and responses make the present way of doing software untenable. "In the past, when we've looked at the job of programming plants, the approach has always

"We engineer our systems to be prepared for every contingency."

been from the top down. We engineer our systems to try to plan for every contingency, then spend many years and many dollars trying to patch and modify the system so it works according to specification. At the end of this process, the system we end up with is nearly impossible to change. It is also brittle."

Emergent systems are those whose behavior cannot be readily predicted from the behavior of their parts.
—George Markowsky

There is a vista of probabilistic algorithms. These might have individual performances that are quite bad, but on the average guarantee good performance.
—George Markowsky

So said George during his talk at the spring '95 "Chaos in Manufacturing" conference, the proceedings for which, dubbed Chaos "3.0" I've been going over in preparation for this fall's Chaos 3.5., to be held in, of all places, Tyngsboro, Mass.

Some of the future conferences will try to tackle issues in social engineering. Rather than just look at the manufacturing processes themselves, it's possible to look at the behavior of the corporate organization as a whole. Going even further, simulation and modeling

systems that are finding some of their first commercial applications in such areas as manufacturing, will eventually be applied to mass democracy as a whole! The behaviors of such beings as politicians will be revealed not to be the actions of individuals, but rather the expression of systemic tensions.

That the king can do no wrong is a necessary and fundamental principle of the English constitution.
—William Blackstone

Think about it. While it may not say everything there is to say about 20th century democracy, it says a lot about how corporations are constituted. More to the point, when it comes to the political realm and emergent systems, is the following:

Without a shepherd, sheep are not a flock.
—Russian proverb

Mechatronics Explained

Tesuro Mori, a Yaskawa Electric executive, coined the term Mechatronics in 1969. The idea is to combine the concepts used in electronics, mechanics, computers and electrics into a single system. This system can be applied to everyday products that themselves are combined applications of these separate disciplines.

For example, Mechatronics applies to cameras, cars, planes, robots, copiers, cellular phones and ATMs. A formal definition might be the "application of embedded complex decision-making to the operation of physical systems." There are many others. Basically, I know one when I see it.

The National Science Foundation (NSF) invited your intrepid columnist to a conference hosted by Tai-Ran Hsu of the San Jose State University. The NSF envisions a global think tank centered around strategies focused on education, products and manufacturing, with Mechatronics as one of the touchstones. Some see the program eventually developing into an international consortium like the Brooking Institute for political studies.

Besides Americans, the attendees were from the Pacific Rim. At the first session, your intrepid columnist had the temerity to question the purpose and objectives of the workshop. I have gotten my wide posterior from attending many such workshops, and too many of them turn out to be "porkshops."

I learned that the real purpose of the conference was to gain respect for the term "mechatronics," which is used very little in America. A few mechanical engineering departments do take a systems approach, but most don't. Consequently, a majority of the conference's purpose was educational.

Who

The real issue at hand, however, is larger than Mechatronics. The problem we all must address isn't a lack of emphasis on Mechatronics, because that's only part of it. What really needs to be emphasized—and even promoted—is education itself.

What

The propensity to give out jackets that recognize, and even celebrate, athletic ability and not intellectual ability is at the core of the issue. If you're bored or rebellious in class, you're told "don't be a wise guy." Engineering enrollment is far behind demand. And what does a practicing engineer end up doing anyway? We attend meetings. We read *Dilbert*. We argue budgets. Living in a cubicle

When

Where

How

For real success in engineering, our education must ensure that at least some of us are unreasonable.

and arguing in meetings is not my idea of creative engineering.

I personally believe that more of the engineering curriculum should be devoted to subjects such as the history of technology (been there, done that); business issues (emphasizing the goal of profit); the who, what, when, where and how of technology (today we only do the "how"); and learning to hear the music. For real success in engineering, our education must ensure that at least some of us are unreasonable. As Shaw stated, all progress depends on unreasonable men.

Perhaps the Mechatronics conferences and workshops will help contribute to that.

The agenda for the next conference was also a point of discussion. Some "classicists" among us wanted to be sure and talk about product development, human resources, manufacturing development and services/maintenance. Luckily, several of the people attending have actually been actively working in the 1990s so we insisted that a few modern topics, like enterprise systems and execution systems, also be discussed.

On the other hand, there was considerable talk about the Internet, and we all promised to link our home pages.

The next conference will be in Japan, perhaps in February, with a theme of "Globalized Local Manufacturing Network for Mechatronics."

Robots and Cyborgs

I recently was asked by Yaskawa Electric (Motoman), one of the premier suppliers of robot technology worldwide, to write down my thoughts about what robotics would be like in the 21st century. I thought I would share some of my random speculations with you.

First of all, robots can be thought of as the culmination of all our technological efforts. Tools are extensions of our hands. Vehicles are extensions of our feet. Cameras are extensions of our eyes. Tape recorders are extensions of our ears. And so on. But a fully-realized robot would be all of those things put together, a kind of super-clone, based on silicon rather than carbon.

Versatile robotics cannot be achieved by means of top-down, command and control architectures. Increasing complexity cannot be dealt with by means of increasing lines of code. What's needed is a bottom-up approach that imparts to the robotic elements in a given system the robustness of an independent agent. It's the change from a state-dependent machine to a sentient capability. This would allow robust performance in the presence of unanticipated stimuli.

A concept related to robots is that of the cyborg, which blends the human with the machine. The term robotics was first used by Isaac Asimov in a story he wrote in 1942. A cyborg, or cybernetic organism, has devices embedded in the body to alter and regulate certain bodily functions. The concept was first described in 1961 as a possible means to enable humans to endure interminable space flights.

Of course, most robots and most cyborgs are not fully realized, but rather devoted to specific tasks. The NC milling machine is one of the most successful machines ever developed. The backhoe in my garden, which we are learning to control over the internet, is a kind of robot. Some people might consider a man on a bicycle as a cyborg.

Our conception of what a robot is has changed, from *Star Wars'* R2D2 to Microsoft's Bob. Just as the idea of the "lights-out" factory has been replaced by trends toward improving productivity though information-technology support of human intelligence, the most important robots of the future may be

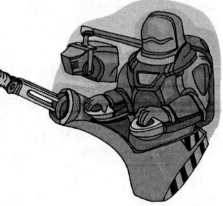

Our conception of what a robot is has changed, from Star Wars' R2D2 to MicroSoft's Bob.

models of ourselves that live on the network.

Robots will be buried in the infrastructure of the future. Virtual robots will browse data, knowledge and wisdom bases. Management and scheduling of assets in natural resources, manufacturing, communications and transportation will enable some of the greatest wealth generation to emerge in the next century. Tiny computers will be everywhere, drawing our baths, cooking our breakfasts and monitoring our excretions.

How entranced shall we be? How deep do the prosthetics go? Are smart wheelchairs okay? Will the lines between carbon forms and silicon forms blur? A building will not be a collection of objects, but an integrated system for comfort, status and safety—a kind of robot.

The big questions are: What will happen when robots free themselves from their anthropomorphic roots? When are they no longer extensions of human physiology, but something truly independent?

For the present, the most viable use of robots is in industry, to automate dangerous or stultifying work. The use of robots to replace humans in other kinds of work is sometimes problematic. However, automation will continue.

I end with a quote from Asimov, taken from the *Handbook of Industrial Robotics*:

"I see silicon-intelligence (robots) that can manipulate numbers with incredible speed and precision and that can perform operations tirelessly and with perfect reproducibility; and I see carbon-intelligence (human beings) that can apply intuition, insight and imagination to the solution of problems on the basis of what would seem insufficient data to a robot. I see the former building the foundations of a new, and unimaginably better society than any we have ever experienced; and I see the latter building the superstructure, with a creative fantasy we dare not picture now."

Part VI — Global Reports

Global competition is now a fact of life. American companies were slow to develop global markets. Constructive relationships have developed over the past decade, even within the same multi-national.

American engineering colleges should offer a second language capability with some geographic seasoning. One only has to visit a Hannover Fair to see how well the Europeans accommodate other languages and cultures.

"Science is always wrong –

it never solves a problem

without creating ten

more"

—George Bernard Shaw

Strengths and Weaknesses

his is the first in a series of personal technology notes from Dick Morley, CEO of Flavors Technology, Amherst, New Hampshire. We encourage readers to contact Dick with suggestions for future columns.

Mary Emrich, the "retired" executive editor, invited me here and I never could refuse a charming woman's request. The content will be technical trends and the future; views will be as personal as possible. It's an honor to be here.

Certainly one of the impacts on technology is the trend towards globalization of the marketplace. No manufacturer wants to build product that can only serve a single, narrow geographic niche. But how can we, as techies, put into our products the invisible cultural features and benefits that will guarantee success in the world market? And can we suppress the all-too-common American "Not Invented Here" syndrome to take advantage of the technologies developed elsewhere in the world?

In 1988, I did a rump Delphi survey to ascertain future directions in the control industry. An update was done in 1990. Results were interesting to say the least. Many surprises. For instance:

- Japan will de-emphasize manufacturing and focus on finance. Japan already has the majority of the world's top ten banks.
- The Pacific Rim will continue to dominate world markets in the manufacture of electronic products for consumer and industrial needs.
- The narrow view of software, the main added value in the newer electronic products, will erode some in the higher-value items. The North American Block will continue to dominate software and high-value electronics in the near future.
- India will become a significant supplier of software. The subcontinent is already a major supplier of such products to many U.S. firms. Its government's emphasis is in software and supercomputers.
- According to the first version of the Delphi study, European "consolidation" was to have a significant impact on machine tool and systems integration. This opinion changed in later updates of the study. The dominance of the new Germany coupled with the long history of fractured geography and language will, according to the pundits,

No manufacturer wants to build products that can only serve a single, narrow geographic niche.

"unbalance" the European initiative.

- A long history of technical excellence in precision machine tools coupled with a well-trained work force will allow a resurgence of tool exports from the Balkans and other countries in the "Warsaw" pact.
- The U.S. will become an integrated part of the "North American Block" along with Canada and Mexico. High-tech products with a huge labor content will continue to be the dominant offerings. Examples includes military hardware, commercial jets, large computers, education and, of course, software of all kinds.

What does this mean to the designer and implementer of manufacturing systems in the good ol' USA? It means that we should play into our strengths and not our weaknesses. Any joint effort with offshore partners should be done with an eye towards the respective strengths of each and not done to correct weaknesses.

Let me cite an example. Recently the Russians approached U.S. enterprises in order to mate the expertise of Russian machine tools and advanced American controls into a viable leading-edge market offering. This is an obvious match in strengths. While the effort has

many difficulties and may never happen, it clearly deserves a strong effort and thorough investigation.

Another joint effort possibility is in the availability of the Indian software expertise at low cost and high quality. Generally the work should be well-specified and have a well-thought-out performance criteria. Using their resources relieves the U.S. company's staff to pursue the more advanced and investigatory aspects of the software in the proposed products.

The lack of interest in the investment community for forages into manufacturing technologies and, as usual, the short-term view of the U.S. financial market opens the door for foreign investment, primarily Japanese. Again, the collaboration of Japanese money and U.S. advanced system know-how takes advantage of each others' strengths.

We should take stock of the situation and work in our areas of strength, not weaknesses. We should not try to catch up but insist that others follow our leadership. As Americans, we should prepare for the future and try not to compete with the past.

Japan and Generation X

First, let me introduce myself. My name is Bob Morley and I've just turned 30 and I am not taking it well. I've been working on high-tech products since I was old enough to solder. I've worked for Functional Automation, Modicon, Flavors Technology and Eloquent Systems, to name a few. I already have a couple of start-up companies under my belt, with about a fifty-fifty success rate. I've also done quite bit of work with the Japanese.

Yaskawa, the Motoman company, sent some of its employees to work at Flavors Technology for nothing. Considering our budget at the time, this was a great boon. But I believe Japan's culture inhibits quality software generation.

The proof? Of the five software engineers that Yaskawa sent us, I got to know two well. The engineers worked hard, and were loyal and committed. However, the Japanese seem to have layers of management for every facet of life, even when playing splatball. The highest-ranking Yaskawa employee ruled the other employees, even outside the workplace. For splatball, the others would never lead or take the initiative. As a result, we would take out the leader and the rest were easy targets.

They are also very concerned with "saving face."

The Japanese I've known never want to be put in the position of saying NO. They think it is rude. For instance, if an American needs a ride somewhere, he'll say, "Hey, can you give me a lift?" If you wanted a ride from a Japanese, you might be wise to start with, "I don't have my car." Then mention, "I had better get started, it's a long walk," and so on, until, finally, he would offer you a ride. It leads to very wordy interactions.

In written agreements and conversations, as well, saving face comes to bear. The Japanese don't actually say no. Instead, they have two responses: a weak affirmative and a strong YES. Any company getting involved in Asian markets has to understand this. At Flavors, we had a product line fail due to just this type of miscommunication. It cost us dearly.

If you want to have a good working relationship with the Japanese, my suggestion is not to rely on paper. Go and find out what's happening. Do exchange pro-

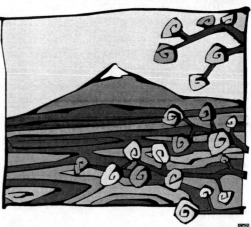

Our columnist's son shares his insights: coming of age in a global marketplace.

grams, a way of trading information they are quite enthusiastic about. They'll learn a lot, but you will, too.

It's important to address the cultural gap. We were a small start-up company at Flavors, so we all had multiple job roles. While designing GUIs (graphical user interfaces), I also made a video for the express purpose of hitting up Yaskawa for venture capital. We translated the audio into Japanese, with the correct dialect and the correct "posture." We had great success, counted in the millions, and we did it by making sure we were in tune with their culture.

So what do they want from us? They need software. They need innovation. Something that we can supply. They are always in search of something to refine. Lately, it's products with the latest chips and techniques, like fuzzy sets and penta chips. Then they say, "We want standards." It confused me at first: Why would standards in communications, designs and protocols be so desirable? Because it makes it easy to refine products. They are hungry for anything that can help them refine techniques and make them more responsive, agile manufacturers. The Japanese excel at refinement. Even their language is a refined version of Chinese.

It always puzzles me when the Japanese are viewed as the enemy. Yes, if you are lazy and overconfident, you have to worry. But they have lots to offer, and they are superb businessmen. I have encountered many American venture capitalists and find that they have to be catered to and sold. The money, if it ever comes, is long and hard. Frankly, the Japanese aren't like that. Maybe they have more faith in American ingenuity and innovation than we do.

So what does the future hold? I suspect that Japan is going the way of California. Its economic climate is not good, to say the least. The new generation wants a living standard like ours, and is willing to sacrifice its heritage to get it. The work ethic is deteriorating, and the Korean and the Chinese are turning into better competitors. They are learning, as America is, that you have to swim harder every day to stay above water. Communication can be tough, but it's a must. Our livelihood is entwined with that of everyone else. The world is getting too small for bigotry and fear of the unknown.

The Oracle Speaks

I act as chairman on behalf of a survey being done to determine which technologies are crucial to the future of U.S. manufacturing. The Delphi survey is meant to ascertain which U.S. technology should be shielded and which should be shared with the rest of world.

The chair first compiles a list of experts. Typically, no more than 50 are chosen. Questions are presented to these experts. (The questions are designed by the chair to be non-leading. This is impossible, but we do our best.)

A short time later, the compiled survey results are put before the same experts. They then seek to "agree" as to the reasonableness of the findings. Of course, the whole process is riddled with subjectivity, and my mathematician friends complain mightily about the process methodology. But it seems to work pretty good.

In any case, we proceed. Early acoustic communications with some of the experts indicated that they fall into four camps. We labeled them "consensus," "offshore," "laymen" and "futurists." The same set of questions goes to each group, and we consolidated the results for each tendency and for the group as a whole. A sample of questions runs as follows:

- What technologies are needed worldwide?
- What technologies must your country protect?
- What technologies must your country acquire?
- What critical technologies will need large-scale cooperation to be successful?
- How can the U.S. help your country prosper?
- How can you help the U.S.?
- How can the American national labs help you?
- What is your company's most important technology?
- What technologies are critical to manufacturing over the next decade?

The consensus group reflected the views of the popular press. America is supreme in manufacturing technology and needs little, if any, help from others. The need to protect ourselves from others intent on stealing "our" technology is of primary importance. Some of "today's" technologies frequently mentioned were total quality management, sensors, software and CAD. "Future" technologies for this group include space-age materials, parametric design and autonomous agents. A xenophobic attitude seems to

What works to our advantage is the ability to move with agility...

dominate this group's responses.

The offshore group, composed mostly of German and Japanese experts, seemed more intent on citing narrow, specialized areas. Their "today" list included fuzzy sets, neurons, JIT and vision. Their "future" list looks to standards as the primary technological need, i.e., "You can't hit a moving target." Their responses suggest a belief that standards help America and penalize others. "Future" issues mentioned include ultra-speed machine tools, ceramics, sensors and software. Notable lacking in an appreciation for systems integration and systems software.

The futurists hold opinions much different than those of the consensus group. The futurists group includes representatives of National Center for Manufacturing Sciences (NCMS), academia and users. Many opined that America has a poor competitive position in manufacturing technologies, and that we have little to shield from others and much to learn. What they believe worked to our advantage was our ability to move quickly and with agility. Key "today" are software, smart chips, industrial computers and innovation. The "future" holds out the promise of autonomous agents, virtual reality, genetic software and neural nets.

My favorite group is the laymen. They are the most difficult to survey, often pleading ignorance or time constraints. We had to resort to face-to-face interviews, and results are fascinating. According to this small but significant group, America has no "today" issues—these were settled long ago. The "future" needs of both America and the world are only two: intelligent systems and nuclear power.

What surprised me most in the results were the discrepancy between the consensus group and the futurists, and the idea that standards threaten America's position. My summary of the results would be: America has software, systems and innovation knowledge; our destiny will not be determined by single-point technologies, but by our agility.

For further information on the USA21 study, contact Charles Hudson at CHudson@ncms.org or fax (313) 995-1150.

Far Out, Far East

I used up about a jillion frequent-flyer coupons buying my wife's airline ticket for our trip to Japan. The main reason for the trip was to visit some old friends at Yaskawa Electric, a company better known in the U.S. as the maker of the Motoman robot.

Among other things, Yaskawa's engineers are involved in emergent-systems programming for the Bullet train, called "Shinkansen" in its native land. Carrying up to 1,300 passengers, the train's top speed is 270 kilometers-per hour—or 167 miles-per-hour—for the metrically challenged. The train moves by means of a 25-kilovolt distributed electrical drive, i.e., no locomotive. Using 3,500 cars servicing 40 stops, the frequency of service during peak hours is every five minutes.

The programming issue involves questions about how to reconfigure the train schedule to take into account unexpected events such as high winds, equipment problems or heavy traffic. Data sampling along the track occurs every three seconds. The rail line's control engineers need to deal with unexpected situations quickly using a software platform that will also be able to handle future use-demands. The life of the control engineer is much like that of a firefighter: long periods of boredom, punctuated with moments of panic.

The future will see increasing speeds—to 350 kilometers-per-hour—with station service frequency of every three minutes. The goal is more speed and better service, while maintaining an excellent safety record. Rules-based, object-oriented software is the designated means for achieving that.

The Maglev, a 550 kilometer-per-hour magnetically levitated linear motor system, using superconducting magnets is also under development.

The bullet train lived up to its advance billing. The creature comforts are great—better than any first-class aircraft cabin I've ever been in. Reserved seating with plenty of leg room, and real quiet. One of stops was in the city of Kyoto, where there are hundreds of religious shrines. With my wife, I visited shrines in the old capital grounds, a kimono show and a demonstration of an original system for punched-card control of weaving operations.

Part of my visit entailed representing EG&G Technology Access Partners, a venture-capital fund. The goal of the fund is to capitalize on technology

It wasn't until after several hours, when I suggested having some beers, that things loosened up.

developed by the American national labs and their 100,000 engineers and scientists. As an advisor to the fund, I was able to visit several Japanese companies and meet with executives. Mitsubishi, for example, was most impressive. I was also impressed with my hosts' understanding of the role technology plays in the creation of wealth and new markets.

Japanese business culture can be difficult. The host sits quietly while the guest makes a presentation. The "audience" may nod, but there's little real agreement or disagreement. In fact, there's little feedback at all. At the end of the bullet-train ride, I gave a presentation at a Yaskawa software facility. But I never could ascertain the level of understanding of the audience. It wasn't until after several hours, when I suggested having some beers, that things loosened up.

I've been in Japan many times, but it was the first time for my wife. We took a bus tour of Tokyo one morning, and she visited a Japanese home. Her impressions of Japan were that it is just like Chicago, except for the language, and that it is very expensive—$25 for a coffee shop breakfast, $100 for a coffee-shop dinner. She thought the people friendly and that they dress well.

I agree with her, although I spent most of my time in business meetings. I found the focus on technology and innovation refreshing, i.e., no MBA talk. The bubble economy has burst and the yen/dollar ratio hovers around 100-to-1, but companies I visited were still more interested in attacking rather than retrenching. Most are more interested in the next five years than in the next financial quarter. Profits in the next quarter are, after all, derived from decisions made five years ago. Decisions made today bear mainly on the year 2000.

The economic problems resulting from the burst bubble, however, are severe. There is pressure to reduce manufacturing costs, and a definite trend toward "build-to order." So-called "salary men" are under pressure to work even harder. Although we, as Americans, will benefit from the Japanese short-term problems, they will be back.

Future Redux

My wife and I just returned from vacation—touring industrial sites in Korea. To give us time for jet lag recovery, our hosts took us on tours of historical sites and temples. After awhile, though, it began to seem as if Shirley and I were the "exhibits," and the stuff in the glass cases was just so much stuff. Kids and others came over to practice their English, and when they learned I was 63 years old they bowed *real low*. I don't get that kind of respect in New York City.

The food was excellent, and the people were warm and gracious. They seem, however, to have an ambivalent attitude toward the United States of America: on the one hand, a grudging admiration; on the other, a resentment of what they see as American post-colonial colonialism. It seems they think some Americans think they know everything. In my experience, true enough.

Anyway, there I was at the top of a monster skyscraper in Seoul and someone is going on to me about how America seeks to export its democratic ideals into other historical contexts and how those ideals may not be right for those contexts. Then I look out the window at the skyline of a city of 10-million-plus people, a city that experienced depredations at the hands of both the Japanese and the North Koreans and is now a bustling, thriving maelstrom of Berlinesque activity.

I'm thinking myself that South Korea could have done worse than imitate the example of the United States, with its largest city's architecture, its free-market capitalism and its form of government. North Korea, which imitated Stalin's Soviet Union, is slowly starving. The United States is not wrong to propagate its ideals around the world. In fact, there are a few places in the U.S. that need to be thus propagated.

While in Korea, I gave a number of lectures regarding my views concerning the changes that will be forthcoming over the next decade and into the next century. Some of the things I talked about already have been explored in this column: the end of mass production, point-of-sale manufacturing, everything on the Internet, the end of software as an art form and artificial life.

As expected, there were numerous nay-sayers in the audience. Students seemed eager to learn, but the old guys sat with arms folded in poses that suggested they were almost

Looking to the past reveals the future of manufacturing.

physically striving to suppress the very notion of change.

That's kind of a silly thing to do considering that a selection of things invented within my lifetime could run as follows: television, penicillin, polio shots, frozen foods, copy machines, plastic, contact lenses, Frisbees, "The Pill," radar, credit cards, split atoms, laser beams, ball point pens, pantyhose, dishwashers, clothes dryers, electric blankets, air conditioners, the technology to put men on the moon, FM radio, word processors, McDonalds, hydraulic brakes and MBAs.

These Korean elders, however, would hardly be the first to fail to see the future staring them in the face. Consider the following:

"This 'telephone' has too many shortcomings to be seriously considered as a means of communication. The device is inherently of no value to us."
— Western Union
internal memo, 1876

"Heavier-than-air flying machines are impossible."
— Lord Kelvin, 1895

"Everything that can be invented has been invented."
— C.H. Duell, Commissioner,
U.S. Office of Patents, 1899

"I think there is a world market for maybe five computers."
— Thomas Watson, 1943

"Computers in the future may weigh no more than 1.5 tons."
— Popular Mechanics, 1949

"We don't like their sound, and guitar music is on the way out."
— Decca Recording Co.
rejecting the Beatles, 1962

"I have traveled the length and breadth of this country and talked with the best people, and I can assure you that data processing is a fad that won't last out the year."
— Editor for Prentice Hall, 1968

"There is no reason anyone would want a computer in their home."
— Ken Olson, 1977

"Hey, 640K ought to be enough for anybody."
— Bill Gates, 1981

Dick Down Under

I just returned from vacation—touring industrial sites in Australia. An e-mail associate, Ian Cook, liked the jib cut of these columns and had me invited. I spoke to audiences in several cities about future technology for manufacturing ("Smaller in Sydney?").

My wife and I flew from New Hampshire to L.A. and then on to Sydney without an overnight, via New Zealand Air, including about 14 hours across the Pacific, non-stop. It's the longest over-water flight in the world, and all together the journey took about 24 hours of no-smoking, aluminum tube travel.

It was our first time in Australia, and we had set aside time for seeing the country. Australia is "only" 200 years old, so we did not have to see any temples. You may remember, in Korea, I OD'd on temples. Here I saw the new #6 blast furnace and a hot strip mill at BHP. From this columnist's perspective, this was one hundred times better than innumerable temples and arty artifacts. Your columnist is not interested in carbon, but is interested in silicon and iron.

Australia is a modern industrialized country. I can tell. I could drink the water and the hotel room had a data plug for my laptop—the real metrics of culture.

I gave two main presentations of three hours each. The audience consisted mainly of engineering and management people from the Sydney and Melbourne areas. They seemed more passive than American audiences I've addressed. More suits and with a formal air. Because my normal audience is rambunctious, I mistook the formality for a lack of interest. Not so. Afterwards, we had good jam sessions.

Many of the attendees felt that while their standard of living is excellent, the country is falling behind in worldwide economic competition. A big concern is the depletion of resources. While other Asian economies, the so-called "tigers" of the Pacific Basin, have few natural resources, they're nevertheless growing like crazy. They create wealth by means of technology alone.

Australia does not do this. Its natural resources are immense, but growth is limited. Australia seems to violate the investment rule, "never spend capital." Assets, it is true, don't do any good lying underground. At the same

Subjects the audiences found interesting were venture capital, chaos and my Harley-Davidson.

time, how can the country thrive when it ships iron ore to Korea, Korea makes cars from the iron, and then ships the iron, now in the form of a car, back to Australia at a profit? It puts Australia in the capital depletion business.

Compared to the U.S., Australia has half of the R&D expenditures per capita and almost twice the tax rate. Doesn't that mean the Australian government "fines" people for being innovative or for doing research and development?

There also is an anti-intellectual bias even greater than that in this country. A group of us were talking about the Olympics to be held in Sydney in the year 2001. The group was lamenting the lack of NBC coverage for the Aussie medals. Proud were they of the muscular achievements of their young adults, and promised vehemently to show the world that Australia has just begun to fight.

The same group then went on to lament the lack of drive in the corporate and academic communities, blaming the economic structure, the unions and the Japanese. My wife suggested grass-roots efforts to change the culture vis-à-vis academic achievement. Something like the bumper stickers seen all over the USA: "My daughter is an honor student at such and such." The silence was deafening. The

guests said that NO Aussie would ever put such a sticker on his bumper, and would be affronted by such a sticker on another car. We were astonished by their contrasting attitudes towards sports and academics.

Subjects the audiences seemed to find especially interesting were venture capital, chaos, lifestyles in the Americas and my Harley-Davidson. Subjects of concern included downsizing, population issues, bureaucracy and the upcoming Olympics. No one seemed very interested in manufacturing execution systems, knowledge databases, or, in fact, software of any kind. This is yet another indication that they lag in acknowledging it is, in fact, knowledge that creates wealth.

We enjoyed ourselves and promised to return next year. The best people there want to create wealth for their grandchildren. And they can do it. But the 50 percent of the labor force that is non-productive will break the backs of the creators, unless Australia becomes a tiger.

Letter From Beijing

I missed the peak of New Hampshire's fall foliage this year because I was traveling in China as the guest of Groupe Schneider, i.e., Modicon Asia, a PLC manufacturer. It took me 29 hours, door-to-door, to go from my house to the hotel in Beijing, without incident. That's a long time to be seated within an aluminum tube. As the airliner descended late one evening upon the capital of the ancient kingdom, the city of 12 million people below did not look much different than any midwestern plains city in the U.S. In other words, for such a huge population center, the amount of illumination seemed limited.

The reason the technical conference was held in Beijing was because it's difficult for some of the Beijing Schneider representatives to leave the country. It's not illegal, only bureaucratically challenging. When we first thought of meeting in China, the Chinese government told us that to do so, we'd have to write letters to President Clinton stating our support for China having certain kinds of most-favored status. We refused to do this and, in addition, threatened to move the conference to Singapore. They caved, and we kept the conference in Beijing.

Once ensconced in our hotel rooms, we opted for the melatonin and sunlight cure for jet lag, and took off to tour Tianmen Square. In the city, bicycles and cellular phones are everywhere. The Chinese, starting from an earlier era of industrial infrastructure, are leapfrogging past modernity into the postmodern industrial age via the microprocessor. No copper wire or telephone poles—it's satellites instead.

Still, the strain of double-digit development on the existing infrastructure is palpable. Several of us visited an American-Chinese joint-venture enterprise that is bent on bringing mass-market amounts of U.S.-style chocolate candy to one of the world's last bastions of communism. The enterprise only gets an allotment of five day's worth of power per week. During the regularly scheduled blackouts, the enterprise would prefer to keep running. So it brought in motor-generator sets for self-generation of power. This sounds reasonable, but where do they get the diesel fuel to run the generators? There are no gas stations. So the fuel

China: a curious blend of the early industrial and post-modern eras.

needs to be shipped in by tanker truck. But there are no tanker trucks, and for that matter, no roads that can handle tanker trucks.

The meetings continued endlessly. My talk went over like a lead balloon. I really bombed. It's a reaction I've noticed before, but American audiences at least are impolite enough to tell me they don't understand me, leaving me some hope of making amends. Next time I give a talk in China I'll be more politically astute and statesmanlike. I need to spend more time on the required pleasantries and not just leap into the technology and speculative thought stuff.

We went to see the Great Wall. They took us to a section where it was guaranteed that there would be no other tourists. It was a two-hour drive, and once we got there it was clear why we were left alone. First, there was a chair-lift ride halfway up, and then an up-and-down, winding climb that was serious enough to call for occasional searches for alternative pathways. My fear of heights also came into play.

Most of the wall—which is one of a few man-made objects that can be seen from outer space—is built atop a line of cliffs. They say that no invading army has ever breached it, but that generals involved in civil wars have upon occasion let the invading hordes in so as to further their own selfish ends.

In general, I found China a rather gray place. And I can't help but attribute that to the leftover communist ideology which places all decision-making in the hands of the state. Many people seemed to spend most of their time leaning on their shovels. Lots of fields that looked tillable had nothing growing in them. Nobody seemed fixated with replacing broken windows panes, coming up with market plans or making a quick buck.

Even though China is such a different place, our meetings revealed that the problems faced by PLC vendors in China, Malaysia, Singapore and Thailand are the same as those faced in Cleveland. These are sales, distribution, costs inventory, service, quirky customers and the like. The same lament. No difference at all.

Delft; Or Virtual Water

Recently, my sojourns into the cyber wilds—where the industrial revolution and the information revolution meet—have taken some spectacularly strange twists. For instance, I was recently asked by Euroseminars to give the opening talk at a drinking water conference in Delft, Holland. They chose me to give said talk in the hopes I'd stir the languid neurons of the typical Dutch control engineer with visions of the future of water processing control. Thus, I found myself on holiday, touring industrial sites in the Netherlands.

My qualifications for speaking about water are minimal. However, my "farm" has some of the cleanest water in New England. At the bottom of the hill is a well field for pumping and shipping bottled water. Most of the water used in the Netherlands, on the other hand, comes from the Rhine River. One of the odd things about water treatment is that the product is of a quality suitable for drinking, yet its main use is in agriculture and industry.

Water treatment is not very different from any other manufacturing process. Multiple raw sources are available, from sparkling well water to waste and sea water. Costs and quantity are an asset allocation issue. Water treatment entails the same process and control issues as all

process production, and as a marketable commodity, raises issues similar to those in the oil industry. Wells must be drilled, pumps run and pipelines (highway and other) operated for final distribution to the user—at roughly the same price as gasoline. Matching the source assets to user demand is one of the biggest problems. Let's face it, geography is one of the oldest truths. Other problems involve purity and conservation.

As with other basic make-and-distribute businesses, time to market and process quality are key issues. The water-to-population ratio diverges sharply from geographic region to geographic region. Seventy-seven percent of this precious resource is used for agriculture and industrial needs, and we deliver drinking-quality water to meet all these needs.

Initially, the group wanted to hear about fuzzy sets and advanced PLC technology. And ignore issues touching on chaos theory. Hmmmm! So we had to ease into the really meaty topic of system complexity. Silo technologies—e.g., PLCs—are well established and have served us well

Where the industrial and information revolutions meet.

for control of processes, in the good 'ol USA and worldwide. Connecting them together via supervisory control, execution systems and the web seems to be the next big step.

My strategy, vis-à-vis our conversation, was to reduce my audience's mental blocks regarding future concepts, not by talking about the future, but instead by explaining the obvious. The process plant, whether semiconductor, paper, water or chemical, will contain green-field control systems that will be valued at 20 percent of the entire facility. The short-term winners are R/3, agents, Microsoft Windows NT, customers, satellites, ADSL, Java, *Dilbert* and the web.

As for system concepts, we emphasized the bulletin board pull-through outsourcing approach. In this manner, much as with the "push" agents subject of so much current discussion, user-supplier relationships will be automatically optimized with modern enabling software. The auto-ERP servers envisioned will have the ability to match, much like dating services, market needs to market supply.

Water found in wells, rivers, the ocean and waste will be offered as sources for drinking-quality water, albeit at different bid currencies. These "bid" currencies will consid-er more than simply raw cost. Product quality, speed of delivery response, reliability of delivery and distance to user are some of these currencies. Relating these sources to processes such as reverse osmosis, filtering, charcoal, distillation and virtual water all will be optimized over the world wide web to meet the needs of society.

Looking over my audience, it was at this point I detected a rustle of interest emerging from their blanket of indifference. "Virtual water? What is virtual water?" Well! Certainly, it is conservation. Water savings garnered from micro porcelain seats and saver shower heads are expected to reach about 30 percent of what's currently being used for such purposes by 2020.

What are some other virtual water sources? It takes about 1,000 tons of conventional agriculture water use to make a single ton of grain. And about 20 tons of grain to make a ton of beef. So, by importing its beef, a society situated where water is scarce can save 20,000 tons of water per ton of meat. Don't use scarce local water resources to supply products that can be gotten elsewhere. Therefore, Holland does well to export tulips, Delft art, civil engineering expertise and other local unfair advantages. And leave the beef to Argentina.

Part VII — Evaluating Entrepreneurship

Complexities of a new business are formidable, and the trend towards starting with a team of complementing skills is nearly a necessity. Many ambitious persons and teams enter the new company path. Few demonstrate the combined technical and business skills to a successful startup.

Successful entrepreneurs evolve from company innovators to corporate tycoons. However, the innovative dynamics in startups is compromised as companies grow.

R. Morley Inc. has been evaluating the characteristics of larger companies. What management methods work best in selecting new ideas? How can transition times be minimized in getting concepts to manufactured products?

"MBAs know everything

but understand nothing"

—*Lee Iococca*

Where Angels Fear to Tread

This is written for those who want to break out of their job and go to work. Worldwide, most work for themselves and few have jobs. But in the developed world, the reverse is true. Yet many jobholders want to escape from the cloistered world of *Dilbert* and the moronic procedural/behavioral rules. In other words, we want to start our own company.

I'm an angel—a venture capitalist—if you will. My group of angels is known as the Breakfast Club. Most angels have guidelines for investment. Ours, for example, say we don't do deals that are more than an hour's drive from home.

We invest in technology, and the earlier the investment, the better we like it. We're looking for 100% market share in all of our investments. This requires serious thought.

The Breakfast Club has four members, with specialists in marketing, finance and the techie—me. We see about a hundred business plans a year, and only invest in several. Once every five years we get to harvest one of our investments. There is no average investment—only failures, the living dead and the winners.

So what interests us? And what can you do to attract an angel? First, you need a business plan. Short

plans are best, about 20 pages. It should cover all of the four-legged dog. The four legs are people, market, dollars and product.

Don't forget to show how we, the investors, will improve our lifestyle as a result of investing in your idea. We want to get back 10 times our investment in five years. We're not interested in a percentage of a failure.

Why do you need us? There are other sources of capital in the early years. Family is best. Borrow and don't pay back. Families will forgive. Don't pay the vendors right away. Make deals with the customers to prepay. In other words, deal, deal and deal. But don't make promises that you can't keep, and stay away from the back pages of *Byte* magazine.

Some say it's hard to start your own company. But what's hard is getting out of business, with the money. Never forget that's the goal. Most liquidity comes from a corporate buyout. Investors' problems come with defining an exit strategy, if any. Management is seduced by the perks and

Many jobholders want to escape from the cloistered world of Dilbert *and moronic procedural/behavioral rules.*

the power, and the stockholders are left languishing on the vine. Make sure the business plan takes into account that the investors want their money back.

Take an electronics hardware manufacturer as an example. About 20 percent of revenue is earmarked for marketing, 10 percent each for engineering and administration, 40 percent for goods and 20 percent is gross profit. Too many disregard the latter and measure success only as revenue. We talk about a $10 million company based on revenue figures. But companies are actually valued at one times revenue or 10 times profit, whichever is lower. Value is more than that, of course. But never trade profit for revenue growth.

What's our batting average? Most venture capitalists are happy if 15 percent of the portfolio succeeds. Return on investment ranges from 10 percent to as high as 40 percent, with the median about 20 percent.

As an investor, we must keep stepping up to the plate. No play, no batting average. Our prejudices are many.

We don't invest in a deal if:

• There is a rubber plant in the front office;

• All of the founders have MBAs;
• The president has a PhD;
• The founder wants to keep absolute control;
• The president drives a Mercedes.

We like it if:

• Investor needs are satisfied;
• There is strong evidence of customer interest;
• There is market dominance;
• There are three founders, plus or minus one;
• Smart people are involved.

Loose Logic for Fanatical Financiers

Recently, I talked to the past president of one of my old companies. We were lucky and the company (Andover Controls) made it. He bought me an expensive steak and told me his sad tale. He tried his hand in venture capital (VC) at the "Angel" level, and seemed not to be doing as well as his peers. Most VC guys have hovered around 20 percent compounded ROI (Return On Investment) in real dollars since the time of Caesar. Most VCs consider the business of investing in small companies as low risk over time.

George was trying to cut his losses by concentrating on lowering risk across the breadth of the portfolio. And in doing so was decreasing the value of the entire portfolio. Why was that so? And why was he dead wrong in trying to reduce risk?

Imagine, if you will, a 20-animal horse race. Each horse has a 1-in-20 chance of winning. And the payoff is $100 for each dollar bet. Clearly we should bet on every horse—our winnings are guaranteed. It would be folly to bet on only one, and if it was losing, to bet even more on the same horse. Yet in many companies, we insist that the

losers be supported at the expense of the winners.

Economists have observed that tax revenues are dependent on the tax rate. So far, so good. If the rate is zero, for example, the state revenue is zero. Conversely, if the rate is 100 percent, the revenues approach zero again. Somewhere between is the optimum tax rate for maximizing state revenue. About 30 percent seems to be the optimum across many societies. If you increase the rate to increase revenue, and the rate is already high, the state net revenue will go down. Fascinating.

In an investment or R&D portfolio, an examination of the extremes leads to the same types of considerations. Having 20 investments of minimal risk (such as government bonds) yields a return of about three percent, discounting inflation. Close enough to zero for me. In fact, if others (competition) are doing the same, you only stay even.

In the same way, infinite risk also yields an R&D benefit of near zero. What is the optimum? About an 80 percent failure rate will optimize portfolio return. In other words, you should do

Sometimes failure is the surest way to success.

deals that have a likelihood of failure greater than half. George has the problem that he tried to reduce risk, rather than increase risk. Risk management is not the unilateral reduction of risk, but the optimization of the risk-to-reward ratio.

In the Darwinian universe, mutations guarantee the continuation of life. Most mutations don't work. But the ones that do are fortuitously adapted to an environment in flux. We cannot predict the outcome of a single coin toss, but we can predict the outcome of a large number of tosses. Having a feel for the "Laws of Large Numbers" is essential for managing portfolios. Many managers are too conservative and manage too closely.

Imagine that you have $20 million to invest. Where do you put it? In a portfolio of 20 companies that have a 50 percent chance of being worth $2 million for each $1 million invested, and a 50 percent chance of being worth $1 million?. Or, in a portfolio of 95 percent chance of zero return and a five percent chance of returning one billion dollars. At the end of the 10 years, you have about $35 million in the first case for your $20 million investment. And in the second case you have a 64 percent chance of becoming a billionaire. Think Bill Gates. The greater the risk, the greater the reward.

R&D investment planning should be centered around failure. Increase risk and the likelihood of success increases across the board. But not for the individual project. We cannot predict individual stars or bums, but we can predict the group behavior.

It is reasonable to establish risk as a parameter of management. But too often the culture of companies penalizes failure for each endeavor, rather than the total success. Success in poker is knowing when to fold, not when to bet. Establish a failure rate and *know* that it's the way to maximize return. Without unpredictability, you lose control and lower benefits. Too much individual risk never happens in real companies. Encouraging failure as a policy seems strange, but it is the way to succeed with minimum risk. Mature companies cannot grow in market share without knowing the "Laws of Large Numbers."

Nonlinear Social Behavior

Oft times this columnist has campaigned for creation of wealth through technology and people's ability to deploy same. My background in physics makes this a no-brainer for me. Less often do I treat the social forces behind emergent technology. But knowledge deployment is the full partner of technology in the creation of wealth.

And wealth is more than "bottom-line" dollars. It can be a quality, an ability or an experience. Companies are formed and evolve as cultural adaptations to historical change. How can we best deal with the DNA of the "Dilbert Syndrome"? How can we create and direct efficient organizations? We can use "top-down" directives, or the "bottom-up" approach. A simple story might help clarify our thoughts on this matter.

Imagine a cocktail party. The problem is how best to manage a party that can exist in two modes: convivial chatter or raucous roar. Conversations are conducted at a decibel level sufficient to be heard over the noise level of the other imbibers. You must be loud enough to be heard, and total loudness is proportional to the number of people in the room.

In one scenario, guests arrive singly. Once there are 30 guests, the tone of the party changes from convivial chatter to a raucous roar, with dancing. At the same time the content of the conversations change from information exchange to something of a much more fundamental emotional context.

In the second scenario, the party never gets to the threshold level of 30 participants. Instead, the host engages in a bit of social engineering—i.e., plays the stereo real loud. This alone pushes the party over the edge from convivial to raucous. If the stereo is for some reason turned off, the noise level of the party-goers remains high. To bring it back down, it's necessary to propose a toast or break a martini glass. The party-goers have no control over the issue. For any two of them to have a quiet bit of conversation it's necessary for them to step out on the balcony.

This simple behavioral model mirrors that for some "emergent" social behaviors, and demonstrates that simple intervention can change complex social states. Individuals have little freedom in the

Crank up the stereo, dim the lights, serve industrial-strength drinks and recruit the first dancers.

matter; you can't command guests to dance or chatter. All you can do is foster an environment where talking is difficult and dancing inevitable. In other words, crank up the stereo, dim the lights, serve industrial-strength drinks and recruit the first dancers.

Why is Morley talking about parties? Is this important? You bet your bippy it is. We've learned the following:

- There's a difference between individual dynamics and emergent social behavior.
- Simple organizations exhibit multiple modes of behavior.
- Behavior modes "lock in."
- Command and control doesn't work.
- Individuals can't have a direct impact on group behavior.
- Altering the environment forces adaptations.

Taking it further, we have a framework for industrial organizations. We can take old behaviors from an old environment and trade them for new behavior in a new environment. To do so we have to remember the following:

- Again, no command and control. Instead, manage the infrastructure.
- Set the rules of engagement and interaction.
- Destroy old sources of individual reinforcements and encourage new reinforcements.
- Establish champions for the new behaviors.
- Reward change-making.

This philosophy can change our way of thinking just as surely as changing salespeoples' commission rates will change what and how they sell.

This article is based in part on a talk given by Chris Meyer of the Center for Business Innovation, Ernst and Young, New York, at the third annual *Chaos Conference* held in Santa Fe, New Mexico.

More About Failure

ast April, I wrote a column entitled "Loose Logic for Fanatical Financiers" which seemed to strike a chord with a number of readers. I argued that failure leads to success, since research and development is susceptible to the laws of large numbers. A zero-risk endeavor cannot return value. Very high risks are prone to failure. We must therefore choose a risk or failure rate most likely to increase the value of our R&D investments.

It's like this: A suggested failure rate for personal bank loans is two percent, and buyers remorse for television hovers around 20 percent. Venture capitalists feel a failure rate of 80 percent is about right to obtain a total portfolio growth of 20 percent per year over a decade, based on the few projects that do succeed. Thus, a failure rate is chosen that leads to overall success.

In like manner, the suggested target for R&D failure is about 50 percent. Choose each project for a 50/50 likelihood of failure. To pursue the "no fail" policy now in fashion will yield the equivalent of Czarist bonds in your stock portfolio. Small wonder large companies have little or no real output for the immense dollars spent on what they call R&D.

Recently, I was fooling around with some software programs for formal evaluation of the process of evaluating R&D programs. It was suggested that the guidelines for the innovation segment (about 10 percent of the total evaluation) should include eight bullet points. They are as follows:

- 10x performance
- 7 years
- 3 gurus
- 100 percent of the market
- 5-legged dog
- 3 rounds
- folding is important
- passion

As Einstein is supposed to have said, "I can't help it if you don't believe it." Let's proceed to explain each of the mysterious Morley cues.

10x performance means that one measurable parameter must be delivered with performance improved by a factor of 10. Not a soft metric, but hard. For example, the price the PC must be 10x lower or the performance must be 10x higher. Note that I said one parameter, not several. The rule is

We only agreed so we could go to lunch.

that diffusion over several factors doesn't count. And several items of 10x dooms the project.

Seven years implies a long time. The project must be managed into three-year segments over a five- to nine-year total life span. Nothing can work that targets the next quarter. Even engineering needed to sustain a product offering averages about 18 months' duration.

I attend many a marketing meeting. Inevitably, the first question the marketing people ask is, "How long will the project take?" We have an answer: Every project takes 18 months and a million dollars. Well! The meeting starts at 10 am and we proceed to argue about the dollar amount and time needed. First, my partner and I concede that a 17-month schedule and a reduced budget is doable, but only because the target release coincides with the mighty "Dilbert" convention in Cleveland.

By three in the afternoon we've agreed to finish up in 11 months, 4 days and 7 hours with a budget of only $438,974.89. The marketing people smile and congratulate each other. But as my friend and I saunter down the hall, he turns and says to me, "Those silly twerps actually think they're going to get the project done in 11 months." The truth is, we only agreed to the schedule so we could break for lunch.

Three "gurus" is the number of "smart" people needed for a project. These gurus need to be supported by no more than five persons. Additional people reduce the likelihood of success and increases the time and money involved. A large team needs to be broken up into small teams with clearly defined shippable deliverables. Every project, no matter how large, needs to be broken up into squad patrol size.

Well, that's three of the eight parameters and we've already run out of space for this month. If you want to find out about the five-legged dog, the three rounds and the importance of folding, you'll have to tune in next time.

Think Ahead

n January, this column discussed various criteria needed for success in research and development (R&D) projects. The space limitations of a columnist's single page prevented elaboration on several parameters discussed as being paramount to successful R&D projects.

At the time, I was fooling around with software providing criteria for formal proposal evaluation. It was suggested that the innovation section—which should comprise about 10 percent of a proposal—should be evaluated against eight criteria:

- 10x performance
- 5-legged dog
- 7 years
- 3 rounds
- 3 gurus
- folding is important
- 100 percent of the market
- passion

The first three parameters were discussed in January, but this columnist assumes nothing about reader's memory.

What you intend to bring to a product to market must be 10x better than anything currently available in some single way. Computers improved at least that amount in the last five years, in several significant ways. Microsoft is striving to leverage its current advantage as we enter the age of Internet. Wal-Mart also is being challenged on several fronts. It takes a significant advantage to sustain a lead for even a short time.

Seven years refers to the time commitment necessary to sustain a new product offering. Next-quarter results should have nothing to do with this quarter's R&D commitment.

R&D projects must be managed by small teams with clearly defined goals. Three gurus—the "smart" people—should be supported by no more than five people each. Remember, too many cooks spoil the broth, and they increase the preparation time and grocery bill, too.

From here we move on to this month's new topics, beginning with "100 percent of the market." To belabor the obvious, market dominance is everything. With dominance in market share, the technologist has control—over pricing, margins and growth.

Look at our friend Bill Gates, for example. Economists say wealth is created by

A successful product must meet future needs.

technology and human ingenuity. Perhaps, but control of the market is the supreme indicator of wealth creation. Market control is obtained by carefully defining an exclusive niche and using an unfair advantage.

Every successful research and development project must make use of the "five-legged dog" mnemonic. The five-legged dog comes from the old riddle, "If you can call a dog's tail a leg, how many legs does a dog have?" The answer is "four," because it doesn't make any difference what you call a dog's tail, it still only has four legs. The riddle has nothing to do with the success of research and development projects, it's the only means to remember that there are five pillars that support successful R&D endeavors:

- management;
- marketing;
- production;
- money; and
- technology.

If any of the five are missing, the likelihood of failure is great. I purposely listed technology last. The fact is, technology is the easy part—we know how to make products. Flow manufacturing and information technology enable us to make just about anything, anytime. The tricky part is selling the stuff; we live in a consumption-constrained society.

Consumption is the starting point. In any project, the first thing written is the user brochure, not the technical specifications of the product. The first question asked is, "What do people need?" not, "What should we make?" We can't tolerate the anarchy of a so-called genius who creates a product just for the sake of creating it. We need to be more professional than that.

On Duty

There are six disciplines or virtues that make an engineer or manager a professional in the same sense that an athlete, artist or soldier is considered professional. These six disciplines include 1) physical rectitude, 2) mental forbearance, 3) temperate habits, 4) correct behavior, 5) magic and an ear for the ineffable and 6) dedication.

Physical rectitude. You gotta stay in shape. Normally, I jog several times a week, several miles per time. I'm 64 years old. You should exercise until you can't think about anything except stopping. Chess players prepare for a big match by exercising, just as boxers do. Tuning the power supply for the wet computer is only good maintenance.

Mental forbearance. Think only about the work. It is the only thing worth thinking about. As do artists and scientists. A master stone mason thinks, dreams and works in stone, not wood. Work a problem over and over until it snaps into place. Focus on the calling and ignore the distractions. Don't worry about being politically correct or well balanced by anyone's definition. Create your image in the shadow of someone you admire. Your calling is music that only you can hear.

Temperate habits. Recognize and correct your bad habits. Since every task is different, adapt behavior to the task at hand. The idea is to take risks, make mistakes and experiment. Most of us still take courses or attend seminars. Don't get credentials, get educated. Don't get caught up in technical drool. Stick to the basics and analyze history for the future.

Correct behavior. As the fundamental underpinnings of emergent objects, individual behavior patterns determine overall performance. Some of the rules are: don't brag; it's okay to look bad or foolish; be an egoist; dream; don't fight it; never hit the brakes; never be satisfied; and get over it.

Magic and an ear for the ineffable. You get zero points for just doing a good job. As an NBA coach recently said, "Everyone's got to have an angle." Magic and flash generate confidence in you and the customer. Go fast, be hungry and act as if you're the oldest inspired adolescent in town. Store up your bag of tricks and attack. think of the big athletes, and their hair colors.

Some rules to live by in the conduct of our daily business lives.

Dedication. This is the last of my irreverent advice. Live the trade. Be loyal, not to your company, but to the music. Work with the best tools, whether they be computers or brushes. Eat Ramen noodles and drink Jolt. Don't get comfortable. Find the challenge. Choose your parents well. One of my interview questions for real techies is, "Do you remember your children's birth dates?" If there answer is yes, their credentials are suspect.

Can one be taught, or learn, to be professional? Probably not. You have to have it in the genes. If you got it, let it out. Otherwise, go into marketing.

I conclude with two of the mottoes I live by:

• living well is the best revenge;
• purity of heart is to will one thing.

P.S. As you undoubtedly know, I'm not the first heavy thinker to come up with a list of virtues like this. The ancient Roman Cicero wrote *De Officiis*, or "On Duties," as a series of letters to his son advising him as to the proper way to conduct himself in public life. Being a more practiced philosopher than me, his list of virtues included only three.

In the same year Julius Caesar was assassinated, Cicero wrote:

The necessity to acquire and maintain the circumstances in which men conduct the business of life is the sphere of [wisdom, courage and temperance]. The purpose is to preserve human society and relationships and to allow the distinction and scope of a man's character to reveal itself, as he increases resources and advantages for himself and his dependents, but even more as he begins to understand that such things are not of the highest value. This area also includes discipline and self-control and moderation and similar qualities, in which some action is brought to bear, not merely the activity of the mind. By maintaining a certain moderation and order in the practical affairs of life we shall preserve goodness and decorum.

Final note: *Some of you may have read one or more of our occasional columns on Chaos theory. If you want to learn more about possible applications of Chaos theory in manufacturing, a conference is being held in Santa Fe on April 8 to 11. For more information, go to morley@barn.org.*

Part VIII — We Have Met The Enemy: They Is Us!

The title of this part is paraphrased from an old *Pogo* comic strip to whom we gratefully give due credit. This brief, but far-reaching statement applies to many of us. These columns poke fun at ourselves, emphasizing that we are better off not taking our situations too seriously.

The twelve articles of this part are intended to be more entertaining than instructive. A healthy sense of humor is downright necessary to cope with the exasperating situations innocent "techies" seem to encounter. So, for this part, sit back, relax and find something to laugh about—which may well be the best section in the book.

Articles Comprising Part VIII:

An Engineer's Logbook • In the Public Interest

Let Me Be Your Guide • Don't Fall in Love • The Art of Herding Cats

A Tale of Two Days • The Raven Consultant

Has Technology Taken Over Your Life?

Geography Is Truth...Or, Where the Bears Are

Resistance Is Futile • Happy New Internet Year

The Road Warrior

Definitions:

- *A Born Executive* —

  a person whose father

  owns the business

- *Experience* — the term

  we apply to our

  history of mistakes

- *Economist* — a fortune

  teller with a job

An Engineer's Logbook

December 8, 1991 (Sun.): The wife and I took our cocktails and sat by the window, watching the new snow fall. We have a fresh new start with no relocation, and I've got a new project underway. Am looking forward to Christmas and the New Year.

January 9, 1992 (Thurs.): The engineering staff finished the new specifications for the customer requirements. What a relief! The spec is short and the design is simple and beautiful. The customer likes it. This is what it's is all about.

February 12 (Wed.): Even though the project is already under way, Marketing wants "minor" changes. And Manufacturing is complaining about costs. As professionals, we must respond by editing our specifications and engineering project plans.

March 13 (Fri.): Upper management reviewed the project. We must redo the specifications to conform with new requirements from various standards committees. Marketing also took this opportunity to change several features.

April 15 (Wed.): Hired a new compiler expert. The software load will increase the power needs. Memory space is scarce. My wife is becoming annoyed at the time spent in the shop. Weekends and nights are spent doing design work and I'm in meetings all day.

May 11 (Mon.): Blew my cool today. Later on I apologized. All that was hurt were my feelings. We have restarted the hardware aspects of the program. The new delivery is 12 months away. Management pointed out that six months have gone by and engineering costs have skyrocketed. Marketing has retained a consultant to help with the project specifications. The manufacturing VP has returned from a long purchasing trip to China and wants a meeting ASAP. Got a little spring fever. Wife is going to spend some time with her mother in the South. Kids and her need the time.

May 29 (Mon.): Just got out of the costing and manufacturing meeting with four of the manufacturing people. The chips we have chosen do not meet with new purchasing standards. Seems like the VP got a great deal of obsolete memory chips in China. We will therefore have to increase the size of the package. This will necessitate a package and heat reevaluation. Have not heard from the wife in several days. Am pretty tired. This is turning into the worst project of my career.

I've been served a restraining order to stay 100 feet away from the marketing guy.

June 18 (Thurs.): Wife and I had a heated argument over the phone until early this morning. She's been trying to call for the last several days. Been sleeping at the office and the secretary blocks the calls if I am in a meeting. Marketing called another damn breakfast meeting.

July 22 (Wed.): My compiler expert quit today. This means a setback in the project schedule. All of us are working half time—12 hours a day. Jerks in manufacturing say we must cut the costs 32 percent.

August 24 (Mon.): Quit today. My secretary, Norma, says I should try to works things out with management. Four messages from my wife. Did not attend the last manufacturing meeting.

August 25 (Tues.): Had a long talk with the suits in the front office. They actually took me to lunch. The talk restored some of my morale. Maybe I went too far. Had two martinis and talked it over with Shirley.

September 25 (Fri.): The wife is putting the kids into a school near her mother's house. And the specs have changed again. If I ever catch that marketing guy in a dark alley, I'll clean his clock. Management has pulled a design review against my wishes. Took a week to prepare for it. The results will not be available for several weeks.

October 26 (Mon.): It rained here in engineering land. Again. The customer review turned on us. The user doesn't want the product the way it is configured. To hell with the suits. We need to get this project under way. Dumped a hot coffee into the manufacturing VP's lap. Felt good. My wife's lawyer called.

November 27 (Fri.): Haven't had a shower for several days. Shirley says that I should take better care of myself. A new project manager has been hired and I am to advise him.

December 15 (Tues.): Bought the kids some nice Christmas presents. Shirley helped pick them out. I've been served a restraining order to stay 100 feet away from the marketing guy. The new engineering manager is very enthusiastic about the new project. Maybe I should hit him, too.

January 20, 1993 (Wed.): Finally quit today. Don't know why I ever wanted to work for this rinky-dink outfit. Am moving to California once we get the final settlement on the house. Rumor has it that the projects out there will keep me and Shirley happy. Looking forward to a new start.

In the Public Interest

Many products today have notices attached to them that alert users to the possible harmful effects of misuse. We believe the same notices should also be attached to many high-tech industrial products. Software, hardware, control systems, maybe even entire plants should have warning labels attached to them. Following is but a sampling of the cautions we think should be considered.

Notice: This plant contains workers. Management lives elsewhere and therefore can accept none of the blame for poor performance. However, they are available to accept the credit for good performance. Any monetary windfalls resulting therefrom should be diverted immediately to management. Enter at your own risk.

Take Heed: This is a project schedule. As all project mangers know, projects invariably take 18 months to complete and entail 40 percent cost overruns. The cost-estimates and schedules contained herein were agreed upon so that the project team could go to lunch. Any resemblance to reality is purely coincidence.

Attention: This disk contains software fully tested by our quality-control department. Users can be confident that adequate precautions were taken to assure error-free product. Our quality is our warranty.

Notice to User: This product is user-friendly. Under no circumstances attempt to operate without the prerequisite three-year training course.

Warning: Contains artificial intelligence. As such, it's smarter than you are. Don't questions its decision.

Alarm: This system is in the database flood plain. At any time, and without notice, severe printer overload can occur. It's recommended that the high-speed printer be connected directly to the system-compatible paper shredder.

Alert: This company is on a downsizing alert. Each employee is on notice that benefits accrued during term of employment are in considerable jeopardy.

Note: Contact with management is best avoided. Engineers are advised to stand at least five feet from any live management representative. If not alive, notify the appropriate administrative assistant.

NOTICE

This plant contains workers. Management lives elsewhere and therefore can accept none of the blame for poor performance.

WARNING: This product contains the infamous NERD virus. Caution is advised.

Advisory: This company generates a significant amount of entrepreneurial spirit, which generally manifests itself by employees walking off with intellectual properties of absolutely no value. Legal fees can escalate sharply as a result.

Safety Alert: Contains state-of-the-art technology. Continued use can cause damage to your career. Don't install unless you plan to jump ship.

Security Alert: A significant portion of this system is porkware. It's added to the Defense Department budget and to the operator's sense of confusion. Approximately 80 percent of the code herein is degenerate. However, that's good, because it's sure to confuse any untutored user.

Caution: The reengineering creed commands that companies reduce themselves to their core competency and concentrate on what they do best. A number of companies that have done so shortly thereafter disappeared.

Prudence: This automation system is controlled by an expert system. As such, even the designers have no idea what's going on. It has great form, but is virtually content-free.

Handle With Extreme Care: This big blue box is a mainframe. Paleontologists agree that its limited memory and low speed enabled immense profit margins. As a result, dominance occurred early in the evolutionary cycle. That is, until smart, fast predators caused the demise of the lumbering beasts. Owners are advised to handle with extreme care. Maintenance costs can escalate beyond measure.

This Side Up: This plant should never be inverted, except by qualified personnel. Personal injury could result.

Danger: This PLC uses ladder logic as its programming language. The tenure of any academic-type found teaching the language will be jeopardized. Electricians are hereby notified that to speak in public of this archaic language is socially unacceptable.

Unsafe: This instruction manual meets all the requirements and specifications you can possibly think of. It guarantees continuous employment for designers and systems integrators. If dropped it can cause considerable damage.

Hazard: This product contains the infamous NERD (New England R&D) virus. Continued use may cause the user to work days at a stretch and consume vast quantities of Coca-Cola. In extreme cases can cause body odor.

Let Me Be Your Guide

Because this column will appear in the Buyers Guide issue of *Manufacturing Systems*, I feel I'm under lots of pressure to produce something that can sit on your bookshelf all year, something you can whip out if whenever you have a tough question that needs answering. It occurs to me that one thing I can do is furnish definitions of some terms you probably won't find in the CIM glossary in the issue's handbook.

Regular readers of this column are aware that we sometimes threaten to sink to new depths in the cynicism and sarcasm department. But that's not the case here. We want to make a serious effort to aid the reader in his or her quest for knowledge. We could simply redefine some of the glossary terms in a vain and feeble attempt at a crude kind of humor. For example, algorithm could refer to the vice president's musical talents. But even we cannot stoop that low.

While we're on the subject of politics, maybe we should pick out some political terms for inclusion in our "Zen" addendum to the CIM glossary. It's appropriate that we give this some thought now, because if the administration ever establishes a proactive industrial policy, the Washington pundits will want to connect our concerns with those of individuals living "inside the Beltway."

Porkware, for example, could be software that answers no market need, but that results from government largess. This phenomenon will of course be written about by professors of porkonomics. And, no doubt, political funding of the data superhighway will be tagged E-Pork.

Another thing we thought about providing is a guide to operating systems and chip technologies. We're convinced, for example, that RISC (Really Integrated Spaghetti Code) will continue to make gains in the marketplace. And Digital's object-oriented language (DROOL) is expected to be announced sometime in 1994. During that same time frame, OS/2 will be upgraded to OS/1.3. Eventually, the system will be simply OS/.

E-Pork Superhighway

E-Pork: Federal funding of the data superhighway.

Did we forget to mention that GUIs (graphical user interfaces) will become even more friendly? Eventually, they will be known as UGUIs (undemanding graphical user interfaces), pronounced "u e gooey."

Because computer chip power will continue to proceed apace, Pentium will be discarded as the *chip de jour*. The name will be purchased by the Tennessee Stove Co. and bestowed upon a local heating element. The superchip of the '90s will be called the Sextium. We can forecast some of the tentative specifications. The clock rate will be 10 GHz with a 128-bit word length. Dissipation in the power-save mode will be less than 340 watts. Systems hardware will utilize the concept of virtual input. This self-generated data will assure the programmer of correct manipulation of input.

Dare we go on? If the reengineering of the American corporation continues much longer, one result is likely to be the emergence of the term "stealth" management. This is because focusing on the core competency of the organization will shrink most corporations to zero. Virtual management and virtual financing will be significant byproducts of this important business process.

Lest people escape unscathed from our glossary addendum, we have acronyms proper to the professions as well. Obviously, a JALAW is Just Another Lawyer and a YAMBA is yet another MBA. And who could fail to discern that when we say "SAPHD," we mean: "Shucks, another PhD."

We hope this discussion has stirred your imagination. If you have any relevant terms or acronyms you'd like to share with us, send them along. But please, send only those that share the kind of serious intent and purpose you find reflected in this column.

Don't Fall in Love

Three notable events fall in February: Presidents Day, Shirley's birthday and Valentines Day. They arrive in sequence on the 12th, 13th and 14th of this, the shortest month of the year—an action-packed month. Politicians, wives and lovers make for strange partnerships, not to mention bedfellows. Our column this month deals with that most basic of teenage passions: Love.

Most of us have a technical background of some sort. We think we are logical and that emotions are absent from our professional life. Alas, not true. All I need to do to evoke an emotional response from many "professionals" is to say that a PC is better than a Mac. Or the reverse. It matters not. Or I can say that software is never complete; ladder logic is better than a flow chart; Intel beats Motorola. Or I can mention any LAN standard. Any or all will be likely to trigger some quirky, idiosyncratic response, borne of equal parts idealism and instinct.

The secret of good design is *good enough*. The enemy of good is *perfect*. The secret of poker is knowing when to fold, not when to bet. A truly successful event is a rare bird. You can't rely simply on the averages when success has such spurious origins. With the Breakfast Club, a group of financial angels I'm

part of, we look at about 100 business plans a year. We invest in about four. About once every five years we hit a winner. Acceptance of that failure rate is the only sure road to success. On the other hand, we have to know when to stop carrying the investment.

Modicon started as project #084 in our venture technology company, Bedford Associates. To this day, every programmable controller Modicon makes has the *nom de guerre* x84 punched somewhere on it. Understand that the significance of the number is that we succeeded with the 84th try. Since then, we've worked on about 400 projects, again, with a winner once every five years. Not a very good batting average. In almost every instance, we fall in love early on and end up hanging on too long. Eventually, however, we quit putting our efforts into a loser and move on to bigger and better things.

What does all this mean in terms of today's automation and information technology systems? Look, for example, at how the increasing imperative for "build-to-

Mr. Underwood missed the boat because he fell in love.

order" rather than "build-to-stock" impacts the thinking of the machine tool builder. The idea that a machine tool (transfer line) can be built around a product belongs to yesteryear. Machining centers will replace transfer lines in the automotive industry. Line manufacturers need to take this into account, making the transfer line of the future act more as a CNC machining center than a build-to-stock production tool.

But the machine-tool manufacturers are in love. They want to believe that the old ways are good enough. They think they're on the right track: they just need to work harder. So they end up making good time, but in the wrong direction. As one of our union leaders stated, "If you always do what you always did, you will always get what you always got."

One of my favorite stories is the tale of Underwood, a manufacturer of typewriters that sold for about $100. Along came the word processor. Sold for thousands. Underwood said "Not to worry." Underwood is no longer in business. Mr. Underwood missed the boat because he was in love.

The same stories can be told regarding the knight in armor; the railroads; the cigarette manufacturers that just want faster machines, incapable of diversity;

and the makers of mechanical printers who have lost market share to laser-printer makers. In fact, the fixed-font, high-speed printer has been bested by the slower, "print-to-order" laser printer. We want a manufacturing facility that has laser-printer attributes.

What can we do? Stay a true professional. If what you need can be bought from a specialist, don't do it yourself. Don't micro-manage. Stay truly logical and ask, "Is this approach valid for today's problems?" The 80/20 rule always applies: 80 percent of the solution can be obtained with 20 percent of the effort. See where you stand and decide whether it's time to fold. The "not invented here" mentality is the biggest enemy of American industry. Remember, as of today, classical automation is obsolete.

Some things deserve love forever. My bride, Shirley, of 35+ years is my love. Couldn't fold if I tried. She raised our three bio kids and about 25 others. She gets gold medals in the giant slalom. She is a good friend, lets me work and I love her. Happy Valentines Day.

The Art of Herding Cats

Managing technical assets is a tough job. Managing software engineers is the toughest of all. The preferred strategy for managing software mavens is to abandon them.

What is software? As the well-known pundit, Ed Kompass, has remarked, "Software is someone else's idea of what you want to do." Most software is written without benefit of insight from the people who will be using it. Its specifcations normally deal with implementation and tool usage, not user needs and goals.

Software is what the machines do, not what the machines are. Yesterday's big concern was hardware implementation. That's because the algorithms of behavior were embedded in iron and copper. Today, we have CNC machines bereft of embedded behavior, which makes them amazingly flexible. The appropriate behavior is software-programmed. Ergo, modern machine tools are process-dependent, not product-dependent.

Just exactly who are the people who write programming code? I are one. We've got posters all over our shop describing the personality characteristics of a composer of factory-floor behaviors. Some of those characteristics are found in the list below. Although humorously presented, you should take these comments seriously. Doing so will allow you, the manager, to better understand the person behind the program.

You might be a nerd if...

- you own two shirts—wear and air;
- your screen saver is X-rated;
- you consider Coca-Cola and Ring-Dings gourmet food;
- your life's dream is faster computers, newer software and more memory;
- your mother cleans your desk at work;
- your dog sleeps under your computer;
- you owe MasterCard more than your annual income;
- Picard, Gates and Seagal are your heroes;
- you don't need a waste basket, since you stand in one;
- you work half-time (12 hours a day);
- your true love is a Macintosh;
- you consider your boss a virtual person;

Do you consider your boss a virtual person?

- you count the days to your next upgrade;
- your personalized license plate says "UNIX-derived";
- you have been fired from a job because of your appearance;
- you mowed your lawn and found a PDP-8;
- you think RAP is "reduced access processing";
- you can find things in the database, but not on your desk;
- *MacWeek* and *Sun-Tzu* are your favorite reads.

So now you have some idea of how we live. For us, access to the building 24-hours-a-day, seven-days-a-week, is a requirement. Free soda and a small refrigerator are musts. We need the latest computers maxed with RAM along with a small budget for applications; a radio and earphones; no dress code; and dogs are allowed. The boss is considered an annoyance, akin to an IRS agent: necessary, but evil.

I've seen software being developed without any specification. Not even a blackboard talk. "The specifications will be written after we do the software." What's a manager to do? Block and tackle.

Insist upon knowing where the effort is headed before starting. Document by means of a design manual. That way the implementation itself is trivial.

Pay adequately, but conservatively. We did not become engineers because the courses were easy or the money was good. We were in love with the art.

Allow total freedom within the goal constraints. Focus on the end product, don't micromanage.

Scheduling is the most difficult thing of all. Early on I noticed that I typically overrun the time allocated for a project by about 40 percent. For the next project I added just that much to my best estimate. But I still overran by about 40 percent. Then I doubled the overrun contingency factor. Once again I overran by, you guessed it, 40 percent. We went back to the original estimate procedure, armed with the vague understanding that we would always be off 40 percent.

As a manager remember these rules:

- Pay enough to satisfy Master-Card;
- Tell them what to build, not how to build it;
- Tell them when they are done.

A Tale of Two Days

It was the best of times, it was the worst of times. So begins the famous novel. In my case, it was the best of times. It was all started when my old friend, Ernie Vahala, suggested he visit me at work. We've known each other some 25 years, and we had yet to meet at our respective lairs. At work to me is a New Hampshire barn.

The idea took off, to the point where the next thing I knew I was inviting the entire Technical Review Board (TRB) of the National Center for Manufacturing Sciences (NCMS) to visit.

The aim was to hold a TRB/NCMS meeting "in the field." It's people that transfer technology to other people. It's done one-on-one and seeing is believing. And there's lots to see in New Hampshire. Modicon manufacturing, for example.

The rules were simple. No ties allowed and beer the way God planned it: from a can. We did allow two choices of beer, 6 oz. and 12 oz. In other words—a field trip.

Well! As my family found out, it's a lot of work hosting 20 "suits." And we had to clean my office. You, gentle reader, may think this is a simple matter. Think more of the 12 tasks imposed upon Hercules by Eurystheus, particularly the one about the stables. We had to clean the outhouse. True. We have about 15 computers and no indoor plumbing. We had to fill the blimp with helium. My son, Bob, took the dump truck to go get the helium. We stocked the cooler and cleaned the week-old take-out Chinese from the office reefer.

The visit's theme began to take shape. We'd tour across time. We'd visit a series of companies that had been started over a 40-year span. Each would prove representative of a level of maturity concomitant with its age. All would be companies I know intimately. The mature 30-year-old is Modicon/AEG. The young adult is Andover Controls. The youngster is Flavors Technology. My barn is the future.

We toured Andover Controls manufacturing facilities on Tuesday. The company is growing so fast its manufacturing system can seem somewhat chaotic. But its products are so good the company enjoys an unfair advantage, and generous margins can offset turbulence borne of rapid growth.

Then we went to Modicon, the 30-year-old company. The people

My son, Bob, took the dump truck to go get the helium.

there treated us right. And by tremendous coincidence, while we were there, a deal was consummated that made Modicon part of a joint venture between Groupe Schnieder and AEG. Talk about customer orientation.

Highlights of the Barn tour that came next included the following:

- a demonstration of an automated micro-chemical sample preparation device;
- a tour of the computer-controlled Barn and a description of the upgrade in progress;
- a look at a model called SimCity and a discussion concerning how complex institutions take a long time to change;
- the start of the blimp project (one of the guys flew it);
- a Flavors demonstration (with support from Ernie Vahala); and
- drinking Harley beer.

We concluded by touring Shirley's house (where I live).

We had a real Yankee supper at Parker's Barn. The food server was appropriately arrogant and the chef was superb. Parker's Barn is much like Morley's Barn. A wood stove for heat. Closed winter and during the week. Dirt parking lot and BYOB. The tables are rough pine boards, with bark, supported on legs of cow yokes.

Parker's Barn manufactures maple syrup the old-fashioned way. They boil it. The visitors therefore saw manufacturing processes dating from the 18th century to the 21st century in a single day.

What did we learn from this intellectual journey across the years? The heterogeneous skill-set found amongst the attendees added to the common knowledge of the group. Some seemed not to believe their eyes, while others immediately suggested improvements. Most helpful were the Texas Instruments and ATT representatives.

Technology transfer takes intestinal fortitude. It needs a change of venue, beer and a casual setting. That happened here, the 19th of April 1994.

The ~~Raven~~ Consultant

O nce within a meeting drea-
ry, while I pondered, weak
and weary,
   Over many a quaint presentation of forgotten lore,—
While I nodded, nearly napping,
   suddenly there came a tapping,
As of someone gently rapping,
   rapping at the meeting door,
"'Tis some consultant," I muttered,
   "tapping at the meeting door:
Only this and nothing more."

Then the sad uncertain rustling of
   his presentation foils
Thrilled me—filled me with fantas-
   tic terrors never known before;
So that now, to still the beating of
   my heart, I stood repeating,
"'Tis some consultant entreating
   entrance at the meeting door:
Some consultant entreating
   entrance on the meeting door:
This it is and nothing more."

Presently my soul grew stronger;
   hesitating then no longer,
"Sir," said I, "or Madam, truly your
   forgiveness I implore;
But the fact is I was napping, until
   so gently you came rapping,
And so faintly you came tapping,
   tapping on the meeting door,
That I scarce was sure I heard
   you"—here I opened wide the
   door—
Hallway there and nothing more.

Deep into that hallway peering, long
   I stood there wondering,
   fearing,
Doubting, dreaming dreams no manager dared dream before;

But the silence was unbroken, and
   the stillness gave no token,
And the only word there spoken was
   the whispered word, "Learn
   more!"
This I whispered, and an echo
   murmured back the word,
   "Learn more!"
Merely this and nothing more.

Back to the meeting turning, the
   carousel of slides still twirling,
Soon again I heard a tapping somewhat louder than before.
"Surely," said I, "surely that is
   something in my software
   lattice;
Let me see, then, what thereat is,
   and this mystery explore;
Perhaps the mailroom and nothing
   more."

Open here I flung the doorway,
   when, with many a strut and
   sway,
In there stepped a high-paid bore,
   he of all the hidden lore.
Not the least obeisance made he;
   not a minute stopped or stayed
   he;
But, with mien of management,
   perched upon meeting floor,

Paymemore!!

Cast a wanton gaze upon my factory
   floor;
Perched and sat, and nothing more.
Then this egoed bird beguiling my
   sad fate into smiling
By the grave stern decorum of the
   countenance he wore,—
"Though thy books be long and
   winded, thou," I said, "art sure
   not craven,
Ghastly grim and ancient soul wan-
   dering from the nether shore:
Tell me what thy lordly name is on
   the consultants' golden shore!"
Quoth the Consultant, "Paymemore."

Much I marveled this ungainly person, to hear discourse so plainly,
Though his answer little meaning—
   little relevancy bore;
For we cannot help agreeing that no
   living human being
Ever yet was blessed with seeing a
   consultant at his door,
Even less to cast a wanton gaze
   upon my factory floor.
With such name as "Paymemore."

But this guru, perching lonely on the
   meeting floor, spoke only
That one word, as if his soul in the
   one word he did outpour.
Nothing further then he uttered,
   not a finger then he fluttered,
Till I scarcely more than muttered,—"Other such ones have
   gone before;
On the morrow he will leave me,
   as all others did before."
Again he said, "Paymemore."

Startled by the stillness broken by
   reply so aptly spoken,
"Doubtless," said I, "what it utters
   is its only stock and store,
Caught from some unhappy client
   whom unmerciful Disaster

Followed fast and followed faster
   till his words one burden bore:
Till the dirges of his soul that
   melancholy burden bore of
   'Pay-paymemore.'"

"Prophet!" said I, "man of evil!"
   Prophet still, if man or devil!
Whether Corporate sent, or
   whether tempest tossed thee
   here ashore,
Desolate, yet all undaunted, in this
   plant enchanted, with its LAN
   much vaunted—
Tell me truly, I implore: Is there—
   is there balm in Windows 4?—
   tell me —tell me, I implore!"
Quoth Consultant, "Paymemore."

"Be that word our sign of parting,
   friend or fiend!" I shrieked, up
   starting:
Get thee back to the schools and
   the hidden shore!
Leave no thick book of that lie thy
   soul hath spoken!
Leave my factory unbroken! Quit
   the seat up on my meeting floor!
Take thy Pen from out my books,
   and take thy gaze off my
   factory floor!"
Quote Consultant, "Paymemore."

And the craven man is sitting, never
   flitting, still is sitting,
In the meeting room just above my
   factory floor;
And his eyes have all the seeming
   of a demon's that is dreaming,
And the praise o'er him streaming
   throws his words upon the floor;
And my soul from out those words
   lies floating on the floor
Shall be lifted—nevermore!

Has Technology Taken Over Your Life?

Most of this was taken off the Internet. Technology has taken over your life if you can relate to the following:

- You think of the gadgets in your office as "friends," but you forget to send your father a birthday card.
- You can no longer sit through an entire movie without having at least one device on your body beep or buzz.
- Your stationery is more cluttered than Warren Beatty's address book. The letterhead lists a fax number, e-mail addresses for two on-line services and your Internet address, which spreads across the breadth of the letterhead and continues on the back. In essence, you have conceded that the first page of any letter you write is letterhead.
- Your network monthly bill is more than your house mortgage.
- You never say excuse me, but instead "service an interrupt."
- At a computer store, you eavesdrop on a salesperson talking with customers Then you butt in to correct him and spend 20 minutes answering the customers' questions, while the salesperson stands by silently, nodding his head.
- You disdain people who use low baud rates.

- You need to fill out a form that must be typewritten, but you can't because there isn't a typewriter in the house—only computers with laser printers.
- You know it's when to have breakfast because sunrise causes you to squint due to screen glare.
- You use the phrase "digital compression" in a conversation without thinking how strange your mouth feels when you say it.
- You sign Christmas cards by putting :-) next to your signature.
- You know what smilies are and can read the following symbols:

  :-) 8:-) :@) :-? :-C ;-)

  :-{) :-D :-o [:-)
- Off the top of your head, you can think of 19 keystroke symbols that are far more clever than :-).
- You constantly find yourself in groups of people to whom you say the phrase "digital compression."

You can relax about one thing ... a computer virus could never survive in this environment..

© CREATIVE MEDIA SERVICES   Box 5955   Berkeley, Ca. 94705

Al Gore strikes you as an *"intriguing fellow."*

Everyone understands what you mean, and you are not surprised that you do not have to explain it.

- You know Bill Gates' e-mail address, but you have to look up your own social security number.
- Your mouse pad is the only clean area on your desk.
- You own a set of itty-bitty screwdrivers and you actually know where they are.
- You think jokes about being unable to program a VCR are stupid.
- When you burn dinner into an unrecognizable mass by trying to multitask, you attempt to restart by pushing "anykey."
- You go to a computer trade show and map out your path of the exhibit hall in advance. But you cannot give someone directions to your house without looking up the street names.
- You know more about Captain Picard than President Clinton.
- On vacation, you are reading a computer manual and turning the pages faster than the others, who are reading John Grisham novels.
- When in an airport, you look for power plugs near the phones. Even though your battery and the backup are at full charge.
- The thought that a CD could refer to finance ro music rarely enters your mind.
- Your definition of a third-world country is one without data jacks.
- You are able to argue persuasively that Ross Perot's phrase "electronic town hall" makes more sense than the term "information superhighway," but you don't because, after all, the man still uses hand-drawn pie charts.
- Al Gore strikes you as an "intriguing" fellow.
- Rather than talk to the person at the next desk, you e-mail over the LAN.
- You have a functioning home copier, fax and computer, but every toaster you own turns bread into charcoal.
- You have ended friendships because of irreconcilably different opinions about which is better—the track ball or the track pad.
- You understand all the jokes in this column. If so, my friend, technology has taken over your life. We suggest, for your own good, that you go lie under a tree and write a haiku. And don't use a laptop.

Geography Is Truth...
Or, Where the Bears Are

Just got off the phone with Dennis Wisnosky. His company, Wizdom Systems, is located in Naperville, Illinois. Dennis is looking to tap into some of the technical labor resources of the northeast. While manufacturing is still the strength of what was once derisively referred to as "the rustbelt," many of the companies that offer software solutions are clustered on either the East or West Coast. Before I get letters, that's not to say that there are no good software companies in the Midwest.

Dennis and I discussed the possibility of managing small groups in several select locations, the primary one being Chicago. It's remarkable how many different skill sets are needed to run a modern software company. And how matching the labor pool's skills to current projects on a day-to-day basis is an even more daunting task. It's a matter of geography, and people.

Dennis and I decided to get a new perspective on personnel management problems using Midwest bear-hunting analogy. In each case, we send different types of technical specialists off to the Midwest on a bear hunt. How they react tells us much about the group's disposition and fitness for meeting the needs of the organization.

As an example, we found that to hunt bears, computer scientists proceed as follows:

1. go to Baton Rouge;
2. work northward, transversing the country from sea to sea;
3. catch every animal seen;
4. 'COMPARE TO A BEAR';
5. 'UPON MATCH, STOP SEARCH.'

Mature programmers place a bear in Chicago to insure that the search will terminate. Taking the analogy just a bit further:

- Assembly programmers prefer to search on hands and knees.
- Mathematicians prove that the probability of catching a bear is too small to be worth their while.
- Professors prove the existence of bears and leave the details to students.
- Engineers hunt bears by catching animals at random using dyna-

Salespeople don't hunt bears, but will sell as yet uncaught bears.

mite, keeping anything that weighs over a thousand pounds and is brown.

- Consultants don't hunt bears, and have never hunted anything. Consultants can, however, be hired by the hour to give advice on bear hunting.
- VPs of technology are hired to hunt bears, but their staff keep the bears safely out of the executive's firing range, thereby ensuring their own job security. If an executive does spot a bear, the staff hires additional people to keep it from happening again. (Senior executive policy assumes that bears are mice with deep voices.)
- Salespeople don't hunt bears, but will sell as yet uncaught bears. Many catch mice and sell them as desktop bears.
- Lawyers don't hunt bears either, but will argue over ownership of leavings. Software lawyers will claim ownership of all bears, based upon the look and feel of a single dropping.

Using the bear-hunt system, Dennis and I should be able to advise people on what their life's work should be, given their proclivities for big game projects. But it's important to keep geography in mind as well. Ask any venture capitalist: each geographic sector has its advantages and advantages.

- *Florida* is a retirement home. Technology centered there seems to languish in the backwater of days gone by. Residents want to only work half a day and have nice boats.
- *New England* is the nerd center of the universe. Entrepreneurs work in their cellars for years, only to emerge and proclaim, "I have built it, and you will buy it!"
- *California* insists that the symptoms of success insure success. Marketing is all.
- *Atlanta* puts up research centers with large staffs.

Years ago, some of us analyzed the American market on the hardware side. The West Coast (including the 51st state, Japan), builds little components—from resistors to computer chips. The East coast (which includes Texas) assembles the components into computers and controls. The Midwest uses the controls to build cars.

To conclude: who and where are key determinants of success. But after discussing all this with Dennis, what did he do? He decided to hunt elephants.

Resistance Is Futile

TO: JIM HEALY
CONSULTANT@MCI-
MAIL.COM
FROM: CLAUS, SANTA
SCLAUS@NPOLE.COM
SUBJECT: REENGINEERING

We have a problem. Our manufacturing system is proving inadequate. Seasonal demand is killing us and we need a way to smooth production despite this variable, unpredictable demand. Read about your services and thought you might be able to help.

****

TO: HEALY
FROM: CLAUS
SUBJECT: SMOOTHING THE FLOW

Your white paper, "CyberSpace Manufacturing: The End of Size as a Competitive Advantage," was enlightening, but probably not useful for the problems at hand.

****

TO: JIM
FROM: SANTA
SUBJECT: TRANSPORTATION

You state, and I quote, "The ordinary 'reindeer-sleigh' delivery system is, at present, a luxury that only wealthy, less-cost-conscious customers can afford." I cannot agree.

We understand the conventional wisdom is we cannot hope to compete with FedEx. But your recommendation that we utilize overnight services and warehouse final assembly runs counter to my staff recommendations. Generally, we vertically manufacture all our discrete entertainment products. We do outsource for semiconductor and software needs.

Are you available for on-site consulting?

****

TO: MR. HEALY
FROM: ST. NICK
SUBJECT: TRAVEL AND FEE STRUCTURE

Wow!!

****

TO: HEALY
FROM: SANTA CLAUS
SUBJECT: YOUR VISIT

We had hoped to impress you with our display of holiday spirit! Alas, as you so cogently point out,

our facility is in the "mature" category. Defined by you, that means "what you've installed, but wish you hadn't."

In particular, you cite our mainframe management tools. We need to, in your parlance, "move to client/server technology with distributed computing across files and applications with PCs and workstations. A grounding in distributed object-oriented computing should be the focus of our efforts. This will eventually lead to simultaneous delivery of concept design and product. This is a cyberspace paradigm in which 1) ideas and products are one, 2) design, production and distribution are separate, 3) tangible goods are limited to enablers and expendables and 4) cyber-product is delivered by means of natural implants."

Thanks again.

****

TO: JIMBO
FROM: BIG SC
SUBJECT: PROGRESS

We've established distribution through the overnight services—eliminating the Pole post office—thereby relegating the reindeer barn to a theme park stop. Both the cookie and gingerbread bakery operations are now outsourced to Keebler, and 30 percent of production in the toy works is now done in Thailand. Your recommendations for going to a non-peak delivery system is under consideration. And I have you to thank for all of this.

****

TO: ABC CONSULTING
FROM: SC FINANCIAL
SUBJ: INVOICING POLICY

Your recent invoice for services was rejected. Only coach travel is permitted. Further, you will not be reimbursed for movies rented during hotel stay. And you need receipts for all purchases over five dollars, including the $7.95 you spent at Barnes & Noble to buy *PCs for Dummies*.

Merry Christmas.

Happy New Internet Year

elcome to 1997. Soon the millennium will be upon us. In preparation for that epoch-marking event, I've begun preparing a list of resolutions and definitions that will define the age we live in. Changes in the technology infrastructure change the language much more radically than you might imagine. Below you will find what I've so far gathered.

In 1997 and in the new millennium, I hereby resolve to:

- always drink Jolt at surfing sessions
- do less ego surfing
- prune my logic trees in the fall
- read more *Dilbert*
- buy new computers before the old ones become chip jewelry
- never use crapplets
- never purchase plug-and-play hardware
- never flame friends—unless they're dead wrong
- never fall asleep giving a talk
- stay away from graybar transactions
- stay away from cobweb sites
- never blow my buffer
- trim my bookmark list
- never use e-mail for wake-up calls
- never use the low-cost solution
- fight implementation standards whenever they float to the top
- become more cash retentive
- clean out my 1988 "in" basket
- stop thinking I need more calculus
- read more S.J. Gould, the paleontologist
- not yell at Karen when she makes quantum physics errors
- upgrade my operating system only once a year
- pay for my software
- buy new gadgets only when half of last years' gadgets are up and running
- teach my Rotweiller to be more sociable
- do expense reports promptly
- always take "the pink stuff" on Asian trips
- realize that the *X-Files* is not a religion
- move my workstation out of my bedroom
- figure out the difference between geeks and nerds

In the new millennium, I will make no resolutions before their time.

- never take the Red Eye again
- buy no companies in 1997
- never purchase Ethernet cards for $1,200
- never use the word paradigm (paradigm—20 cents)
- never use my tongue as a volt-meter
- keep sufficient supply of WD-40 and duct tape on hand
- never calculate the lunch tip on my laptop
- never answer e-mail before coffee
- never roll my eyes on hearing that someone doesn't have e-mail
- never turn repairing household appliances into a control systems project
- have dinner at home at least once a month
- take clothes on trips
- stay the Alpha Geek

The Road Warrior

Most of us are familiar with the standard preparations and activities associated with business travel. We're hip to the non-taxable income opportunities that come with airline bonus mileage. The office can handle the car rentals, hotel bookings and flight reservations. However, to be a better traveler, some urban fables need to be debunked. Tips about some unusual aspects of travel may prove helpful to the novice, intermediate or even expert business traveler. The following advice is for the road warrior:

- Pack for one extra day. I spill airline salad dressing on my pants, so the extra set always helps me right out of the box.
- On the other hand, pack half the clothing and twice the money you think is needed. Go bag lite.
- Put a LARGE colored ribbon on all the luggage.
- In my "kit," I carry a wind-up alarm clock, water purification pills, ibuprofin, Sudafed, melatonin and meds. Spare eyeglasses also are a must.
- Take two bags with handles and rollers, one for clothing—as large as needed—and a carry-on for an on-board Newton, books and other irreplaceable items. In my carry-on, I take a neck pillow, computer, travel paperwork, noise-cancel-ing ear phones, reading materials, pain and nose medications and one set of underwear. Never use garment bags. Learn to pack and fold.
- Always check the big bag. Navigating Chicago's O'Hare airport is tough enough without hauling a large bag. Nevermind the idea of "lost baggage." It will get there— a day late maybe, but it will get there—so check your luggage.
- Get on the plane early. It helps resolve overhead bin crowding, computer foul-ups and upgrades.
- Always fly business class when flights are longer than two hours.
- Lounges are great. Use them for e-mail and non-alcoholic drinks. In the lounge areas, sit near a power outlet. Sometimes you can meet a friend whilst sucking up power for your laptop.
- Speaking of laptops, I recently bought an Apple Newton 2000. I own a dozen home computers now. Why another? Well! The laptop I normally use roasts my leg or melts the plane table. My Newton runs for a week on four

AA cells with intermittent use. (Normally, I take two normal batteries and Bubba, the Godzilla of lead cells, on every trip, yet somehow I'm always out of power.) Yes, it's not a Cray, but I can write, mail and stay cool. The Newton is different. The first week, I almost used it for skeet practice until I realized the problem was me. Now I am a fan.

- Cheap hotels are OK, but big beds are great. When the reservation is made, say you are six feet, four inches tall and need a king-sized bed.
- Hotel food is OK, too. There are three kinds of food in the world. Fuel, social and good. Fuel comes from almost any kitchen. Make sure all the four food groups are ingested every day. These four are: brown, takeout, leftover and frozen. A possible fifth is chemical, to cover the diet drinks.
- Hotel room utilities normally are OK. My definition of a third-world country is one that has hotels in which you should not drink the water, and has no data ports on the phone. The U.K. is a good example.
- Taxis are many times better than rentals, especially in foreign countries. Watch out for directions from natives—they give directions thinking you "know that place where the church used to be."
- Contact home often with e-mail and postcards. I mailed postcards to my mom, now departed, at least once a week, and never was hassled like my brothers with the lament, "at least your brother Dick always writes." The only message on the postcard was "Hi, Mom." I also send cards to the geography class at the local elementary school.
- At the meeting, take lots of notes. Turn off the sound on the laptop to mask game noises. Take naps often.
- Don't forget the flowers—stop to smell the roses. When in Beijing, I was wedded to my laptop. The host arranged a trip to the Great Wall, and I was going to stay in the hotel with my umbilical cord attached. Dumb. Went to the Wall. I do get tired of viewing piles of rocks, but the history perspective never hurts.
- When you return, you'll remember where you parked your car because you always park it in the same place: on the roof, in the corner closest to the elevator or bus pickup. Even thieves avoid rain and snow.
- Send thank-you notes to your hosts.

Remember, there are problems with every trip, so relax.

Index

Richard E. Morley, best known as the father of the programmable logic controller (PLC), is a leading visionary in the field of advanced technological developments. An entrepreneur whose consistent success in the founding of high technology companies has been proven through more than three decades of revolutionary achievement, Mr. Morley has more than 20 U.S. and foreign patents, including the parallel interface machine, hand-held terminal and magnetic thin film. His MIT-based background in physics has provided insight into becoming an internationally recognized pioneer in the areas of computer design, artificial intelligence, automation and futurism.

As an inventor, co-author of *The Technology Machine: How Manufacturing Will Work in the Year 2020*, consultant and engineer, Dick Morley has provided the research and development community with world-changing innovations. His peers have acknowledged his contributions with numerous awards, honors and citations. Mr. Morley's medals of achievements are from such diverse groups as *Inc.* magazine, the Franklin Institute, the Society of Manufacturing Engineers and the Engineering Society of Detroit.